American Institute of Chemical Engineers

Topical Conference On Hydrogen 2006

Held at the 2006 AIChE Spring National Meeting

April 23-27, 2006
Orlando, Florida, USA

Printed from e-media with permission by:

Curran Associates, Inc.
57 Morehouse Lane
Red Hook, NY 12571
www.proceedings.com

ISBN: 978-1-60423-530-2

Some format issues inherent in the e-media version may also appear in this print version.

American Institute of Chemical Engineers

Topical Conference on Hydrogen
2006

TABLE OF CONTENTS

Session 24 - Hydrogen Sensor Development

Cochairs: William Rhodes
Maximilian B. Gorensek

Chair: Jim Brenner

1 **Fiberoptic Hydrogen Leak Detector**
Barry G. Grossman, Sachin N. Dekate

2 **Nanocrystalline Doped Sno2 for Room Temperature Detection of Hydrogen; Recovery, Response, and Sensitivity Improvements**
Christina Drake, Sameer Deshpande, Satyajit Shukla, Linda Nguyen, Divya Bernard, Sudipta Seal

3 **Low-Power Detection of Hydrogen Leakage Using a Self-Powered Wireless Hydrogen Sensor Node**
Jerry Jun, B. Chou, J Lin, X Shengwen, A Phipps, K Ngo, D Johnson, A Kasyap, T Nishida, H. T. Wang, B. S. Kang, F. Ren, L. C. Tien, P. W. Sadik, D. P. Norton, L. F. Voss, S. J. Pearton

13 **Optical Hydrogen Sensor for Inert Environments**
Jerald A. Cole, Robert A. Lieberman, Alexander Trunov

22 **H2 Detection Via Polarography**
Jesus A. Dominguez, Ron Barile

35 **Development of Hydrogen Detection Instruments at Kennedy Space Center**
Brian M. Hunter Jr., C R Arkin, Charles Curley, David Floyd, Guy Naylor

Session 51 - Hydrogen Topical Plenary

CoChair: Sharon Robinson

Chair: Maximilian B. Gorensek

39 **Challenges and Benefits of a Move Towards Hydrogen Fuel**
Steven G. Chalk

40 **An Overview of the Nuclear Hydrogen Initiative**
Carl J. Sink Jr., Amy C. Taylor, Paul S. Pickard

46 **Thermochemical Cycles for the Production of Hydrogen from Water Using Nuclear Energy**
Robert J. Evans

47 **"Sustainability Issues Related to the 'Hydrogen Economy'."**
Earl R. Beaver

Session 99 - Past, Present and Future Uses of Adsorption and Membrane Processes in Hydrogen Production

CoChair: Armin D. Ebner

Chair: James A. Ritter

48 **Chemical Industry Vision 2020 Overview**
Francis Zia

49 **Overview of Hydrogen Production Technologies**
James A. Ritter, Armin D. Ebner

51 **Purifying Hydrogen with Inorganic Silica Membranes at High Temperatures**
Mikel C. Duke, Suraj Gopalakrishnan, Madhat Abdel-jawad, Michael Macrossan, Joe Diniz da Costa

56 **Fast Cycle Rotary Adsorption Bulk Separator at Low Pressure**
Surajit Roy, Bowie Keefer, Aaron Pelman, Andre Boulet, Matt Babicki, Graeme Herring

57 **Hydrogen Production Via Pd-Ag Permeators: Membrane Reactor Design and Process Study**
Silvano Tosti, Angelo Basile, Livio Bettinali, Fabio Borgognoni, Claudio Rizzello

64 **High-Pressure Co2/Ch4 Separation Using Sapo-34 Membranes**
Richard D. Noble, John L. Falconer, Shiguang Li, Janna Martinek, Tracy Q. Gardner

66 **Novel Thermal Swing Sorption Enhanced Reaction (Tsser) Process Concept for Hydrogen Production**
Hugo S. Caram, Alexander Verdooren, Shivaji Sircar

Session 142 - Past, Present and Future Uses of Absorption and Membrane Processes in Hydrogen Production II

CoChair: James Ritter

Chair: Armin D. Ebner

67 **Adsorption and Membrane Processes in Hydrogen Production**
James A. Ritter, Armin D. Ebner

69 **Chemically Crosslinked Polymer Nanocomposite Membranes for Hydrogen Separation**
Christopher Aberg, Jon Samseth, May-Britt Hagg, Richard Spontak

71 **Sorption-Enhanced Reaction Processes: Steam Methane Reforming Combined with in-Situ Co2 Removal for Increased Hydrogen Production**
Alírio E. Rodrigues

74 **Methanol Steam Reforming and Ethanol Steam Reforming in Membrane Reactors: Experimental Study**
Angelo Basile, Adolfo Iulianelli, Fausto Gallucci, Silvano Tosti, Enrico Drioli

79 **Use of Room Temperature Ionic Liquids for Carbon Dioxide Separations**
Richard D. Noble, Carl A. Koval, Douglas Gin, Dean Camper, Jason Bara

81 **Gas Separation in Coal Gasification**
Joe Carlos Diniz da Costa, David S. Sholl, Graham Reed, Kelly Thambimuthu

Session 182 – Hydrogen as an Energy Carrier: High Volume Production Issues I

Cochairs: Sharon Robinson
Timothy R. Armstrong

Chair: Maximilian B. Gorensek

88 **Pre-Conceptual Design and Cost Estimate for a Nuclear Hydrogen Production Plant**
William A. Summers, Maximilian B. Gorensek, Edward T. Danko

90 **Experimental Studies of the Iodine-Water System for the Sulfur-Iodine Thermochemical Cycle**
Jacob M. Crosthwaite, David Bruce, Mark C. Thies

92 **Sulfuric Acid Decomposition Catalysts and Reaction Considerations for Sulfur-Based Thermochemical Water Splitting Cycles**
Daniel M. Ginosar, Harry W. Rollins, Lucia M. Petkovic, Kyle C. Burch

99 **Improvement of the Thermal Efficiency of Hydrogen Iodide Concentration in I-S Process by Using Radiation-Induced Polymerized Membrane in Electrodialysis System**
Hiroyuki Okuda, Tetsuya Yamaki, Shinji Kubo

105 **Development of Separation Processes Based on Membrane Technology for the S-I Thermochemical Cycle**
Giampaolo Caputo, Claudio Felici, Alberto Giaconia, Salvatore Sau

107 **Technical Barriers and Opportunities in Nuclear Plant/Hydrogen Plant Connection Technologies**
Steven R. Sherman

109 **Hybrid Sulfur Cycle Flowsheets for Hydrogen Production from Nuclear Energy**
Maximilian B. Gorensek, William A. Summers, John W. Weidner

110 **Solar Production of Hydrogen Using a Cadmium Based Thermochemical Cycle**
Lloyd C. Brown, Barry W. McQuillan, Gottfried E. Besenbruch

Session 221 - Hydrogen as an Energy Carrier: High Volume Production Issues II

Cochairs: Sharon Robinson
Steve Sherman

Chair: Maximilian B. Gorensek

111 **Nuclear Hydrogen Initiative Calcium-Bromine Cycle: Determine the Feasibility of Using Cold Plasma Dissociation for H_2 Generation, and Molten-Spray Contactor for Ca•Br_2 Regeneration**
Richard D. Doctor, Mike Henry, C.B. Panchal, Yu Po Lin, Steven A. Lottes, Robert W. Lyczkowski

113 **Testing of the Palladium Membrane Reactor and Potential Applications for Hydrogen Production**
William Rhodes, Dave Howard

114 **Logistics Fuel Catalytic Cracking for Hydrogen Generation**
Mukund Karanjikar, James Stevens, Lixin You

115 **Dynamic Simulation of Reforming and Regeneration Processes for Fixed-Bed Adsorption-Enhanced Natural Gas Reformer**
Lixin You, Balaji Krishnamurthy, Yunquan Liu, Farshad Bavarian, Curtis Krause, James Stevens

117 **Hydrogen Generation Using Fixed-Bed Adsorption-Enhanced Natural Gas Reformer**
Balaji Krishnamurthy, Kerry K. Spilker, Lixin You

118 **Analysis of a Novel Process with Integrated CO_2 Capture Using Coal and Thermal Energy from a High-Temperature Nuclear Reactor for the Production of Hydrogen and Electric Power**
George Tsatsaronis, Kerstin Kapanke, Ana Maria Blanco Marigorta

127 **Analysis of a Nuclear-Powered Ammonia Plant**
William L. Kubic Jr., Andrei V. Belooussov

137 **Aspen Plus™ Process Design for the Liquid Hydrogen Production by Steam Reforming of Used Automotive Lubricating Oil**
Mohamed Elbaccouch, Karthikeyan K. Ramasamy, Jong Baik, Ali T-Raissi

Session 256 - Hydrogen Storage Scenarios for Transportation Applications

Cochairs: T. Motyka
Sharon Robinson
Timothy R. Armstrong

Chair: Maximilian B. Gorensek

143 **Solid-State Hydrogen Storage System Development and Engineering Analysis**
T. Motyka, D.L. Anton

145 **Vehicular Hydrogen Storage: Goals, Challenges, and Progress**
Grace Ordaz, John Petrovic, Carole Read, George Thomas, Sunita Satyapal

146 **Bench Scale Saes St 909 Tests for Methane and Carbon Dioxide Removal from Helium, Hydrogen, and Nitrogen Gas Streams**
James E. Klein

147 **Novel Light Weight Complex Hydrides for Hydrogen Storage**
Sesha S. Srinivasan, Elias K Stefanakos

155 **Fundamental Structural Properties of Single-Walled Carbon Nanotubes: Reopening the Debate over Hydrogen Storage**
Sandeep Agnihotri, Yun-Fat Lam

160 **Safe, Efficient Storage of Hydrogen in Liquid Ammonia and Ammonia-Based Solid Chemicals**
John H. Hobrook, Norman G. Olson, Robert K. Graupner, R. Doyle Miller

166 **Emissions from a Hydrogen-Compressed Natural Gas Fueled Ford F-150**
Henry Ng, Robert W. Peters, Fouad H. Fouad

167 **Examination of Thermal Performance of Insulation Systems Combining Novel and Conventional Materials for Lh2 Storage**
Lixing Gu

Session 268 - Hydrogen Purification and Fuel Cell Applications

Cochairs: William Rhodes
Maximilian B. Gorensek
Sharon Robinson
Timothy R. Armstrong

Chair: Jim Brenner

169 **Nanoscale Investigation of Morphologies in Polymer Electrolyte/Pvdf Blend Membranes**
James T. Goldbach, Karren L. More, Michel Foure, Scott R. Gaboury, David Mountz, Amy L. Manheim

179 **Membrane Desulfurization of Logistic Fuels for Fuel Cell Auxiliary Power Units**
Kandaswamy DuraiSwamy, Richard Root Woods, Diane Aagesen

185 **Hydrocarbon Membranes Composed of Poly(Vinyl Alcohol) Derivatives for the Application of Fuel Cell**
Sang Yong Nam, Ji Won Rhim, Hoon Huh, Seong Yong Ha

186 **Development of Dense Cermet Membranes for Hydrogen Separation**
U. (Balu) Balachandran, Tae H. Lee, Ling Chen, Sun-Ju Song, Stephen E. Dorris

187 **Performance of Pd-Alloy Membranes for Hydrogen Separation from Mixed Feed Streams Containing 1000 Ppm H2s**
Osemwengie U. Iyoha, Bryan M. Morreale, Richard P. Killmeyer, Robert M. Enick, Bret H. Howard

192 **Hydrogen Purification with Palladium Alloy Membranes**
Jim Brenner, Gaurav Bhagat

24a Fiberoptic Hydrogen Leak Detector

Barry G. Grossman and Sachin N. Dekate. Electrical Engineering, Florida Institute of Technology, 150 University Blvd., EE Department, Melbourne, FL 32901

Fiber optic sensors for hydrogen leak detection have unique advantages with the potential to address many of the shortcomings of other technologies. Fiber optic sensors have a small form factor and are lightweight. Due to the nature of their operation, they can be deployed in hazardous environments and monitored from a remote location. They are safe in explosive environments and are immune to electromagnetic interference. They can be optimized for any desired parameter-of-interest and in some cases measure multiple parameters simultaneously. Fiber optic hydrogen sensors can be orders of magnitude more sensitive than their counterparts using electrical technologies.

Like many electrical sensors, our approach incorporates palladium as the critical sensing component. Uniquely, we accurately measure the expansion of palladium when exposed to hydrogen and temperature with fiber optic sensors. Since temperature is a major factor in determining the concentration of the hydrogen leak, the ability to measure temperature and hydrogen concentration simultaneously is necessary. In this paper we report on results of our sensor development and testing. This includes tests at 2% hydrogen in nitrogen which result in a measured sensor response time of less than 10 seconds.

24b Nanocrystalline Doped Sno2 for Room Temperature Detection of Hydrogen; Recovery, Response, and Sensitivity Improvements

Christina Drake[1], Sameer Deshpande[2], Satyajit Shukla[2], Linda Nguyen[2], Divya Bernard[2], and Sudipta Seal[2]. (1) Advanced Materials Processing and Analysis Center, University of Central Florida, Engineering Bldg 1, #381, Orlando, FL 32765, (2) AMPAC, University of Central Florida, Engineering Bldg 1, #381, Orlando, FL 32765
Current hydrogen sensor issues are selectivity, use at lower temperatures, and fast detection. Use of nanomaterials in sensors may improve detection limits to lower temperatures by alterations of the space charge layer. The surface reaction which identifies H2 on the sensor is improved when the number of defect sites for reaction are increased .

Nano In-SnO2 is synthesized using sol-gel method with for ppm level H2 sensing at room temperature. Its sensing is based on variations of the depletion layer in the presence of a reducing or oxidizing gas, which leads to a variation in the height of the energy barriers for free charge carriers. Conduction electrons have to overcome a potential barrier induced by the space charge layer. Doping with indium enhances the conductivity response by within the SnO2 lattice. Ceria, due to its multi-valence states, has oxygen storage and release capabilities. It will be incorporated with the In-SnO2. In an oxygen-free environment, CeO2 would release stored oxygen to In-SnO2 to recover its original resistance

Variations in doping amount of indium for increased sensitivity and response are being investigated for generation of excess oxygen-ion vacancy concentration. Doping in the amounts of 1%. 3%, 6.5%, and 9% will give insight to the optimal doping amounts for room temperature sensing using the sol gel method to derive nanocrystalline SnO2. Variations in doping amounts of Ceria are also being investigated. Preliminary results show that room temperature sensing of hydrogen can been achieved using this system. Preliminary results have also shown that doping with Ceria can reduce the recovery time to under a minute, recovering approximately 95% of the sensor's original conductivity.

LOW-POWER DETECTION OF HYDROGEN LEAKAGE USING A SELF-POWERED WIRELESS HYDROGEN SENSOR NODE

J. Jun, B. Chou, J. Lin, A. Phipps, X. Shengwen, K. Ngo, D. Johnson, A. Kasyap, T. Nishida
University of Florida, Department of Electrical and Computer Engineering, Gainesville, Florida 32611

H. T. Wang, B. S. Kang, and F. Ren
University of Florida, Department of Chemical Engineering, Gainesville, Florida 32611

L. C. Tien, P. W. Sadik, D. P. Norton, L. F. Voss, and S. J. Pearton
University of Florida, Department of Materials Science and Engineering, Gainesville, Florida 32611

Abstract

Within the ongoing interdisciplinary hydrogen research at the University of Florida, a self–powered wireless hydrogen sensor node has been designed and developed from a system level approach. By using multi-source energy harvesting circuitry designed and developed at the University of Florida, scavenged or "reclaimed" energy from light emitting and vibrational sources serve as the source of power for commercial low power microcontrollers, amplifiers, and RF transmitters. After system power up, our sensor node is capable of conditioning and deciphering the output of hydrogen sensitive ZnO nano-rods sensors also developed at the University of Florida. Upon the detection of a discernible amount of hydrogen, our system will 'wake' from an idle state to create a wireless data communication link to relay the detection of hydrogen to a central monitoring station. Two modes of operation were considered and designed for the use of hydrogen detection. The first mode would sense for the presence of hydrogen above a set threshold, and alert a central monitoring station of the detection of significant levels of hydrogen. In the second mode of operation, actual hydrogen concentrations starting as low as 10 PPM are relayed to the receiver to track the amount of hydrogen present.

The paper will discuss the performance of our Self-Powered Wireless Hydrogen Sensor Node, and also focus on the design and optimization of the detection circuitry, digital processing considerations, and modulation scheme to maintain an accurate and reliable system while expending a minimal amount of energy scavenged from ambient light or vibration, for long lifetime operation with minimal maintenance.

ZnO Nano-Rod Sensors

The Zinc-Oxide (ZnO) nano-rods developed at the University of Florida between the collaboration of students and faculty of the Department of Materials Science and Engineering and the Department of Chemical Engineering, were used as the hydrogen sensing mechanisms of our system. These lightweight hydrogen sensors were designed with the goal of achieving high sensitivity, rapid response to stimuli, reversibility, and low power consumption, all within a physically small and light package.

The unique characteristics of ZnO nano-tubes and nanorods make them fundamentally appropriate candidates for the sensing of hydrogen. ZnO is a material currently used in the detection of humidity, UV light, and gas, and has shown to change resistance with respect to both temperature and hydrogen exposure. Because of its wide bandgap of 3.2eV, the ease of synthesizing nanostructures, the availability of heterostructures, and the bio-safe characteristics of this material, ZnO is a most attractive material for the specific sensing application at hand[1]. With ZnO nano-rods placed in an array, as a gas sensor, they are able to create a large chemically sensitive surface-to-volume ratio which is needed for high sensitivity in hydrogen sensing. ZnO nano-rods can also be produced cheaply, and are highly compatible with other microelectronic devices.

To enhance the device's sensitivity to hydrogen, catalytic coatings or dopings of platinum (Pt) or palladium (Pd) were used to further increase the ZnO nano-rods hydrogen sensing mechanisms[1]. Figure 1 shows the transient response to different dopings of hydrogen for ZnO nano-rods coated with Pt.

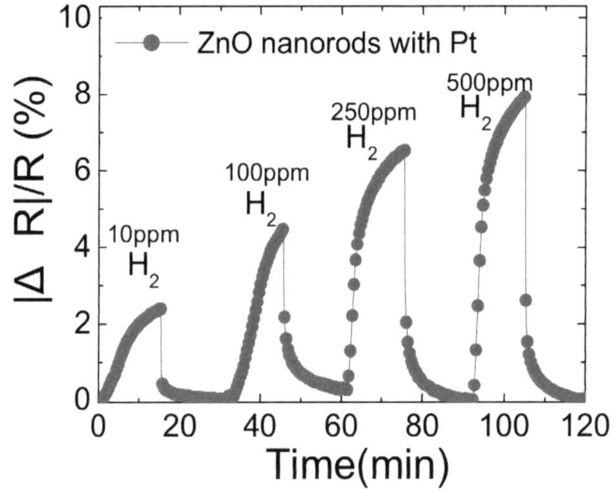

Figure 1: Transient Response of Pt coated ZnO Nano-Rods for different Hydrogen Dopings

Differential Detection Sensor Interface

The challenge in designing the interface between a sensor and the Analog-to-Digital (A/D) converter of a system is found in the necessity to obtain an accurate real world signal with the limitations of low power and reduced voltage swings[3]. Given that the ZnO nano-rod's initial response to any exposure of hydrogen is distinct and immediate, this intrinsic characteristic will serve as an ally for the successful detection of hydrogen.

Since the ZnO nano-rod's resistance changes with respect to how much and how long the device has been exposed to hydrogen, the most popular and accurate way of detecting resistance changes is through the use of a Wheatstone Resistive Bridge. Such a bridge can be seen in Figure 2(a). By using the combination of a Wheatstone Resistive Bridge, and a difference amplifier as seen in Figure 2(b), We can create an extremely low power and portable interface for the detection of hydrogen. This interface can be seen in Figure 3.

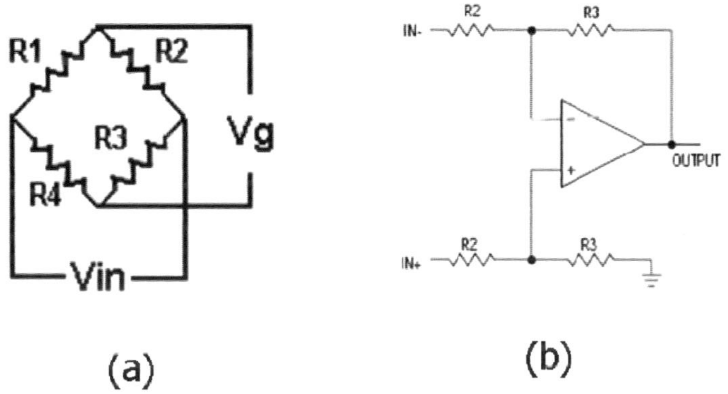

Figure 2 : Wheatstone Resistive Bridge(a) and Difference Amplifier(b)

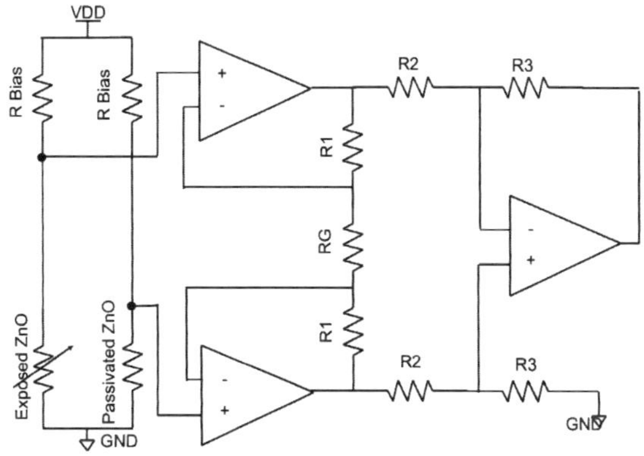

Figure 3 : ZnO Nano-Rod to ADC Differential Detection Interface

Within the diagram of Figure 3, we can see that the Wheatstone Resistive Bridge of Figure 2(a) and the Difference Amplifier of Figure 2(b) were combined through an additional non-inverting gain amplifier stage to buffer, amplify, and provide a high impedance input to the Wheatstone Bridge before the signal from the resistive bridge is processed by the difference amplifier. This topology of amplifiers is also known as an instrumentation amplifier.

In the Wheatstone Resistive Bridge stage, the main objective is to detect the differences in resistance between a ZnO nano-rod which is exposed to hydrogen, and that of a passivated ZnO nano-rod which is encased in glass. By using a passivated ZnO nano-rod encased in glass to be the reference resistor, we can detect only the resistance changes caused by exposure to hydrogen, and not that of those caused by other variables such as temperature. So, when no hydrogen is present, the passivated and exposed ZnO nano-rods should be similar in resistance, and the output voltage of our sensor to ADC interface should be approximately 0 V. We designed our instrumentation amplifier using MAX4289 Op-Amps from Maxim-IC. These Op-Amps were chosen due to their low power requirements (1.0V / 9uA) and for their typical low input offset voltage of 200uV.

By providing a reference voltage of 2V taken from our supply voltage, and using the onboard 10- bit A/D converter of the MSP430, a low power microcontroller, the A/D has a resolution of about 2mV, with 1024 voltage levels between 0 and 2V. The output of the interface must be able to meet this requirement and provide at least a 2mV per ohm (output voltage to ZnO resistance change) output. The measurements for our fabricated sensor interface can be seen in Table 1.

Table 1 : Sensor Interface Measurements

Supply Voltage	ZnO Resistance	PPM	Supply Current	Power	Sensor Output
2V	1565	0	42uA	84uW	80uV
2V	1521	10	44.2uA	88.4uW	138mV
2V	1500	500	44.3uA	88.6uW	240mV

Given the A/D resolution of 2mV, from Table 1 we can see that our sensor and sensor interface is capable of detecting the presence of at least 10 PPM of hydrogen.

Wireless Link

RF Transmitter

The complexity of the modulation was the key component in our selection of a wireless RF front end. We needed to satisfy the RF transmitter's requirement for ultra-low power consumption operation without sacrificing our transmission distance. It was decided that a lower modulation scheme equaled lower transmitter architecture complexity, which meant the ability to realize a RF front end with fewer discrete power consuming components.

The simplest modulation scheme available is that of a "carrier present, carrier absent" technique, also known as "On Off Keying" (OOK). What made OOK an appealing modulation scheme was the intrinsic premise that an OOK transmitter would only be "on" and consuming power when the RF front end was transmitting a "high" or a "1", and that little to no power would be consumed on the transmission of a "low" or "0".

We decided to use a Ming TX-99 transmitter tuned to 300 MHz to serve as the RF front end of our sensor. The architecture of this transmitter is based on a colpitts oscillator design seen in Figure 4 and consists of a single transistor and a LC tank to tune the transmitter to oscillate at a specific frequency. Initial tests show that when biased at 0.6 V, the transmitter drains 850uA, which translates a power requirement of 510uW to transmit a constant 50% duty cycle 580mV peak to peak square pulse train of 1 kHz. We are also currently using a ¼ wave 22 gauge copper monopole antenna tapped at one of the inductors of our LC tank to increase our transmission distance.

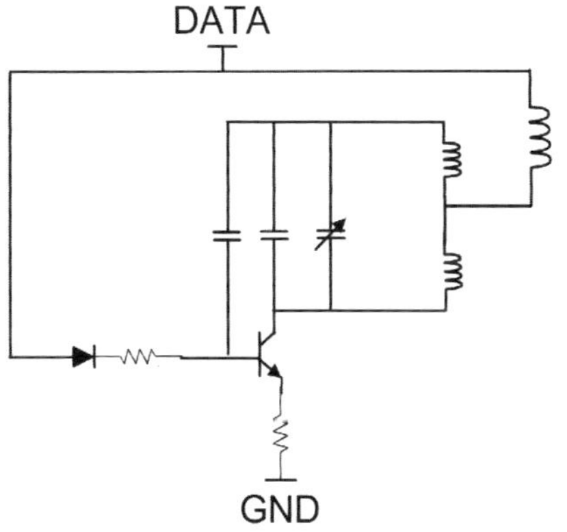

Figure 4 : Transmitter Schematic

Central Monitoring Station

These transmitters will be communicating to a central monitoring station consisting of a receiver, data acquisition device, and a laptop. The receiver is a LC based Ming RE-99 receiver module tuned to 300 MHz. The data output of the RF Module is tied to a Labview USB-6008 data acquisition (DAQ) devices which deciphers the output of the receiver and displays the information onscreen in a gui written in Labview code. Both the DAQ device and the receiver module are powered from the USB port of a laptop, so power consumption is of little worry to the development of the central monitoring station.

Microcontroller

Microcontroller Selection

The proper selection of our microcontroller was essential to the success of our design. We required the microcontroller to include an onboard ADC with enough resolution to track the changes of our sensors and be capable of conditioning and processing the data received from our sensor interface. We also needed enough onboard memory to retain both the runtime code as well as store the data sampled by the ADC. There is also the requirement for our microcontroller to have the ability to encode and send this data to our transmitter via a serial output. Because we are using OOK as our modulation scheme, the system would be optimized if the microcontroller included a serial output port capable of sourcing enough power to drive and power our transmitter. We also required the microcontroller to be easily reprogrammable and consume a minimal amount of power.

Because our goal is for a truly self-powered system, we must place an emphasis on the assumption that the bulk of our power will be consumed in the active states of the microcontroller and transmitter. However, because the time required for the ZnO Nano-rods to saturate is on the order of minutes, our system can operate with a very low duty cycle, and so, idle or standby current also becomes a significant factor in the decision of which microcontroller we select.

We eventually selected Texas Instrument's MSP430F1232IPW. We chose this specific microcontroller because of its many features, large searchable knowledgebase, and the quality of assistance and samples given by TI. Table 2 highlights all the pertinent features of our microcontroller selection.

Table 2 : Features of TI MSP430F1232IPW

Type of Program Memory	Flash
Program Memory	8 kB
RAM	256 Bytes
I/O Pins	22 pins
ADC	10-bit SAR
Interface	1 Hardware SPI or UART, Timer UART
Supply Voltage Range	1.8 V – 3.6 V
Active Mode	200uA @ 1 MHz, 2.2 Vsupply
Standby Mode	0.7 uA
# of Power Saving Modes	5

Modes of Operation

Currently the microcontroller is programmed to run as a state machine, and has two different reprogrammable modes of operation. In each mode of operation, the microcontroller operates within the following states: initialize, collect data, transmit data, and sleep. The first mode of operation is for the level monitoring of hydrogen. This mode runs through each state until a discernable threshold of hydrogen is detected. This threshold is set so that although hydrogen is present, the level of hydrogen is not enough to pose any serious danger. Once hydrogen is detected, the microcontroller forces the RF front-end to transmit an emergency pulse to the central monitoring station, and returns back to an idle mode.

The second mode of operation is of data transmission. In this mode, the microcontroller collects data from the sensor interface, and queues this data to the RF front-end to be transmitted to the central monitoring station. This mode is for a constant tracking of hydrogen levels, while the level monitoring mode is to alert the end user that hydrogen has indeed been detected. The state flow diagram for the Level Monitoring Mode and Data Transmission Mode can be seen in Figure 5.

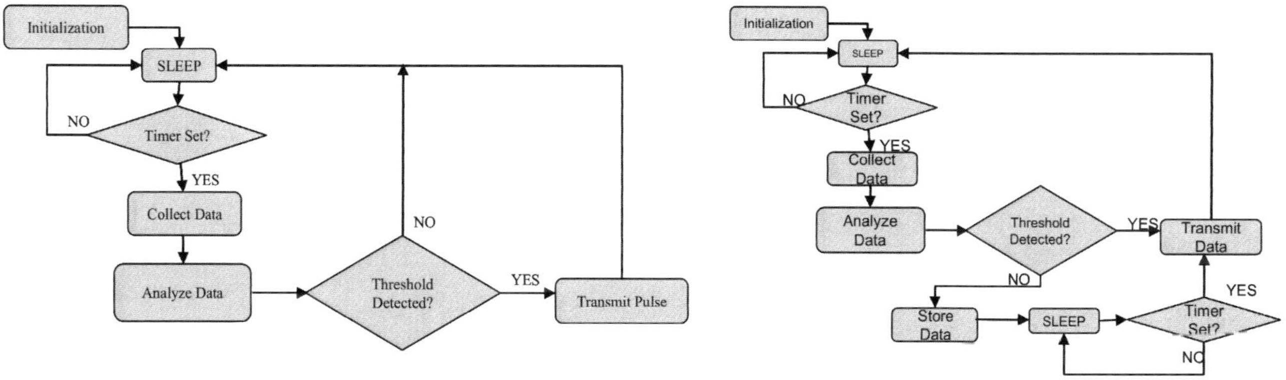

(a) (b)

Figure 5 : Level Monitoring State Machine(a) Data Transmission State Machine(b)

Energy Harvesting

The objectives in the design of our energy harvesting devices involved the ability to extract energy from two main sources of energy—Solar and Vibrational. The end product was one that was able to extract energy from both photovoltaic devices, and piezo-electric (PZT) energy harvesters. This energy would then be primed for use by the hydrogen sensor, and sensor interface, as well as the micro controller and transmitter. A system block of the energy harvesting devices can be seen in Figure 6.

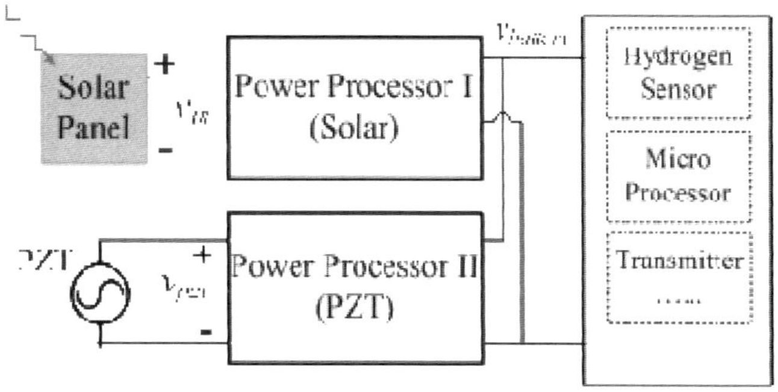

Figure 6 : Energy Harvesting System Diagram

Solar Energy

Photovoltaic devices are a mature commercial product, and offer the availability of high energy density per area. They are however, limited to lighting and temperature conditions. For our system, we chose to use an IXOLAR XOD17-04B Solar Cell as the input to the power IC developed at the University of Florida. This power IC, which was designed and fabricated, can be described as a Pulse Resonant Power Converter designed to be self-powered and self-controlled for maximum power point tracking, low switching loss, and to convert an input voltage of 0.8-1.2V to a steady 2V output voltage which can be used to power the hydrogen sensor, microcontroller, and RF transmitter. The solar power system was tested in lab conditions using a flash light as the energy source, and was successful in delivering a sufficient amount of power to power the microcontroller, sensor interface, and RF transmitter.

Vibrational Energy

For the harvesting of vibrational energy, piezo-electric devices are attractive as sources in that no light is required, and the collection of energy is proportional to the volume of the devices. The limiting factor however, is the magnitude and frequency of the vibrations. As MEMS PZT devices are currently being developed, for a proof of concept design, commercial PZT bimorph beams were purchased and used. We chose to use four PSI D220-A4-203YB Double Quick Mounted Y-Pole Bender as our PZT devices, and connected this to a direct charging circuit (full-bridge rectifier and shunt capacitance) constructed at UF. By adjusting the proof mass on the PZT, we were able to change to resonant frequency of the PZT bimorph device. We tested our vibration energy harvesting system under lab conditions, and used a mechanical shaker tuned to 1 g_{rms} @ 130 Hz, the resonant frequency of the bimorph beam, as our vibration source. This proof of concept design was also able to provide 250 uW to successfully power our hydrogen sensor node

Power Analysis

A power analysis was done to examine the power requirements of our system. From this analysis, we observe that the power consumption for the microcontroller to remain idle, output data via serial power, and to scan the ADC's input, is a constant 2.5 uW. The most power consumed by the microcontroller at any time, is in the microcontroller's initialization state, which occurs only once during initial power up of the microcontroller. The initialization time for the microcontroller is only for 12.5ms, where average power consumption is 3.07mW with a peak power of 7.3mW (as seen in Figure 7).

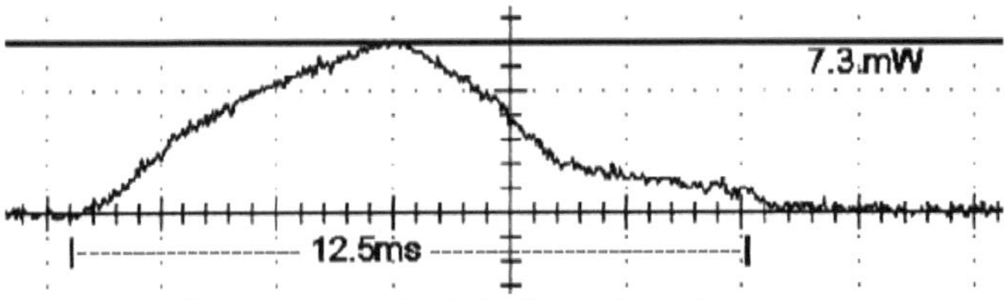

Figure 7 : Initialization Power

With the Ming TX-99 transmitter attached to the serial output port of the microcontroller, we also verified that our RF transmitter module does not consume power for the RF transmission of a logic "0", and that an average power of 261uW with a peak of 522uW (as seen in Figure 8) is consumed for the transmission of a "high" 500us pulse width. The charging and discharging characteristics of Figure 8 is due to the LC resonant tank which is used to set the operating frequency of our RF transmitter to 300 MHz.

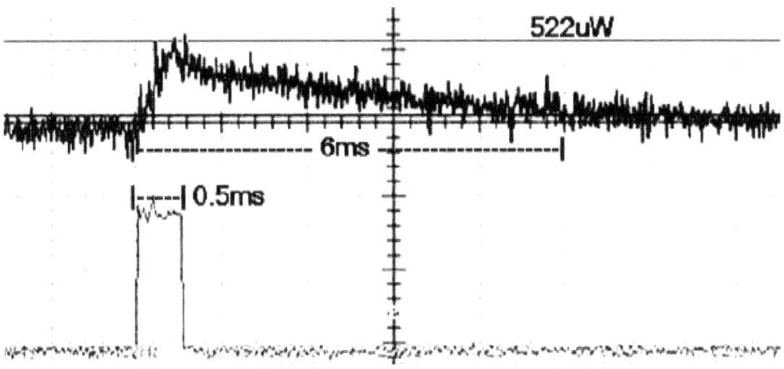

Figure 8 : Power to Send 1 Bit

A power analysis was also done for the sensor interface. This power analysis includes both the power to drive the instrumentation amplifier, as well as to power the biasing circuit for the resistance bridge. From our measurements, we see that when no hydrogen is present, our sensor interface consumes 84uW of power, and for 10ppm and 500ppm, our interface consumes 88.4uW and 88.6uW respectively as detailed in Table 1.

System Integration and Testing

All blocks of our system were integrated and successfully tested under laboratory conditions. Our system is capable of both detecting and transmitting the presence of 500 PPM of hydrogen while obtaining power through energy scavenging techniques. A hydrogen chamber at the University of Florida was used to expose a controlled 500ppm doping of hydrogen to our ZnO nano-rod sensor. We tested our system under the constant data transmission operating mode to see that our sensor interface was indeed showing a change in data pattern as our sensor interface detected the rising amount of hydrogen.

The analog and digital components of our system were tested by using solar and vibrational energy reclamation techniques separately, and were successful in both scenarios. Because a constant data transmission mode has a higher power requirement than the level monitoring operation mode, our testing procedures show that both modes of operation can operate from the energy scavenged through our harvesting techniques.

Summary and Remarks

The design considerations and results of our Self-Powered Wireless Hydrogen Sensor were presented. Our system consists of solar and vibration energy harvesting devices with power management circuits used to power the sensor, sensor interface, microcontroller, and the RF front end of our sensor system. Under laboratory conditions, we were successful in powering and maintaining a wireless link to relay the detection of a controlled 500 PPM hydrogen source. Our results indicate that levels as low as 10 PPM can also be detected.

Current and future work being performed on the self-powered wireless hydrogen sensor include Minimum Energy Coding to further minimize our power consumption for data transmission, and the integration of nano-sensor, detection circuit, microcontroller, RF transmitter, antenna, energy harvesting, and power management circuits into a single compact module. We are also currently in the process of fabricating and testing our MEMS scale PZT devices to be used with a power IC specifically developed for vibration energy harvesting. These would replace the commercial bimorph beams and direct charging circuit to increase the compactness of the overall module.

References

1. L. C. Tien, P. W. Sadik, D. P. Notron, L. F. Voss, S. J. Pearton, H. T. Wang, B. S. Kang, F. Ren, J. Jun, and J. Lin, Appl. Phys. Lett. 87, 222106(2005).

2. Rayming Corporation. "TX-99 V3.0 RF Transmitter Board." Datasheet, Jan 2001.

3. "Making Low Power and Low Voltages Work in Analog Design." Application Note 747. Maxim-IC. 17 Apr. 2001 < http://www.maxim-ic.com/appnotes.cfm/appnote_number/747>

4. Maxim. "MAX4289 1.0V Micropower, SOT23, Operational Amplifier." Datasheet, Jan 2002 < http://pdfserv.maxim-ic.com/en/ds/MAX4289.pdf>

5. Texas Instruments. "MSP430F1232 16-bit Ultra-Low-Power Microcontroller, 8kB Flash, 256B Ram, 10 bit ADC, 1 USART." Datasheet, Jan 2002 < http://focus.ti.com/lit/ds/symlink/msp430f1232.pdf>

6. A. Chandrasekaran et al., "Design considerations for distributed microsensor systems," Custom Integrated Circuits, 1999. Proceedings of the IEEE 1999, p. 279.

7. Y. Prakash and S. K. S. Gupta, "Energy efficient source coding and modulation for wireless applications," in *Proc. IEEE WCNC'03*, vol. 1, New Orleans, LA, USA, Mar. 2003, pp. 212–217.

OPTICAL HYDROGEN SENSOR FOR INERT ENVIRONMENTS

Jerald A. Cole
Hydrogen Ventures
Long Beach, California

Alexander Trunov
Intelligent Optical Systems
Torrance, California

Robert A. Lieberman
Intelligent Optical Systems
Torrance, California

Biography

Jerald Cole is President and Chief Technology Officer of Hydrogen Ventures, a technical and management consulting firm in Southern California. Mr. Cole has over 25 years experience in process R&D, including over 15 year developing advanced hydrogen production and purification technologies. He has held various senior technical and management positions at OptiSense, LLC, Hydrogen Burner Technology, GE Power Systems, and Energy and Environmental Research Corporation. Mr. Cole holds a B.S. (with distinction) in Chemistry from San Jose State University and an S.M. in Chemical Engineering from the Massachusetts Institute of Technology.

Alexander Trunov is Group Leader of the Optoelectronics and Signal Processing Group at Intelligent Optical Systems, a developer of optical sensor technologies in Torrance, CA. He has over 12 years of experience in electronics, communications, and control systems, and specializes in the areas of fiber optic sensing, as well as signal and image processing. Dr. Trunov holds M.S. degrees in Electrical Engineering and Biomedical Engineering from the Polytechnic Institute of Brooklyn, NY, and the University of Science and Engineering, Moscow, Russia, respectively, and a Ph.D. in Electrical Engineering from the University of Cincinnati, Ohio.

Robert Lieberman is President and Chief Technology Officer of Intelligent Optical Systems, a developer of optical and fiberoptic sensor technologies located in Southern California. Dr. Lieberman has over 25 years experience in development of optical technologies and is a Past President of the International Society for Optical Engineering. He has held senior technical and management positions at Physical Optics Corp, Optech Ventures, and AT&T Bell Laboratories. Dr. Lieberman holds B.S. and M.S. degrees in Physics from Rensselaer Polytechnic Institute and a Ph.D. in Physics from The University of Michigan.

Abstract

A hydrogen sensor has been developed based on a modification to the Wacker chemistry used for reversible detection of carbon monoxide. Critical elements of the modified chemistry are metallic palladium stabilized by a heteropoly acid, supported on a porous Vycor™ matrix. In inert environments such as helium or nitrogen the sensor responds over a range of hydrogen concentrations from ppm levels to several percent by volume. The reversible response is measured as a darkening of the sensor within the visible spectrum. This mode of operation lends itself to fiber optic coupling of the sensor to light source and detector. Response times are typically only a few seconds except at very low concentrations, where mass-transfer effects become important. The sensor is fully reversible and recovers rapidly. The behavior of the sensor has been modeled as an equilibrium process between gas-phase hydrogen and two phases of hydrogen-palladium solution. This model permits linearization of the sensor signal over at least three orders of magnitude in hydrogen concentration. In the presence of atmospheric oxygen the sensor exhibits complex behavior for hydrogen concentrations below about 1,000 ppm, suggesting catalytic oxidation of the hydrogen. At higher hydrogen concentrations the response is comparable to that observed in an inert gas environment.

Introduction

Intelligent Optical Systems (IOS) has been developing hydrogen sensor for NASA Stennis and Kennedy Space Center for leak detection in hydrogen-fueled launch vehicles. Nominally, these sensors are expected to operate in a dry, nearly inert environment, which is a challenge for many other sensor designs. In addition, NASA has determined that sensors networked using copper wire present an unnecessary safety risk. As a result, a requirement for this application is that the sensors be fiber optically coupled within the launch vehicle. This, of course, dictates an optical sensor response.

After reviewing prior art a decision was made to adopt Vycor®, a nanoporous glass matrix, to support a chemochromic hydrogen-sensitive dye. The Vycor matrix was fashioned into 3-mm diameter rods that could then be cut to convenient lengths and coupled to an optical sensor using low-cost 1-mm PMMA optical fiber. The resulting sensing elements are known as "optrodes". As a result of their high porosity (*ca.* 30 percent) the optrodes were expected to yield a rapid response time (*ca.* 1 – 3 s). The optrodes were also found to have a wide dynamic range, limited largely by the sensitivity of the light sensor and the stability of the light source.

Background

Hydrogen optrodes based on the palladium-catalyzed reduction of tungsten are often referred to as "Benson-Tracy" sensors (1). There have been many physical embodiments of these sensors (2), but the thin-film fiber optical sensor is one of the most studied. A critical concern with all Benson-Tracy sensors produced to date has been the response time. Typically, even with micron-thin palladium films, response times to step changes in hydrogen concentration are measured on the order of tens of seconds, with even longer recovery times. In the present work, the response time issue was addressed by embedding the sensing chemistry in a Vycor matrix. This reduces the characteristic dimensions of the sensor chemistry from micron dimensions down to nanometers.

Embedding the chemistry in Vycor was done by serial impregnation of first tungsten oxide and then palladium. For example, a tungsten (VI) oxide layer can be readily generated by soaking the Vycor matrix in $W(CO)_6$ in ethanol, drying, and then decomposing the carbonyl under ultraviolet light. Palladium is provided from an aqueous solution of palladium (II) chloride. In the resulting optrode, the palladous ion (Pd^{2+}) undergoes an electron transfer reaction with hydrogen.

$$Pd^{2+} + H_2 \Rightarrow Pd^0 + 2H^+$$

The reduced palladium then reduces W^{6+} to W^{5+}, and returns to Pd^{2+}state.

$$2W^{6+} + Pd^0 \Rightarrow 2W^{5+} + Pd^{2+}$$

This optrode formulation has a response time of about 1 – 2 s, depending on the hydrogen concentration, and is completely reversible. A blue color can result as the optrode darkens, because the change in transmissivity is more pronounced in the yellow than in the blue end of the spectrum. We found that without Pd^{2+}, (i.e. with metallic palladium) the reduction of W^{6+} to $W^{5+}$is extremely sluggish.

Reversibility of the hydrogen optrode requires that W^{5+} be reoxidized to W^{6+}. This requires an electron acceptor (a Lewis acid). The fact that we have observed reversibility indicates that Vycor contains electron-trapping holes. Vycor glass, indeed, contains ca. 2 percent (w/w) boron. Boron is an electron deficient element ready to oxidize W^{5+} to W^{6+}. We have not explored how reduced boron would dissipate its newly gained electron, but presumably this occurs through production of trapped water, as a result of stoichiometric reaction between solid-state oxygen and protons.

In the earliest work conducted here (3) we used W^{6+} and Pd^{2+} in stoichiometric proportions. However, we saw an aging effect due to this excess of Pd^{2+}, where Pd^O once formed, was partially oxidized to PdO through reaction with atmospheric oxygen. Presumably this occurred because much of the palladium was electronically decoupled from the tungsten. Regardless, the Pd^{2+} catalyst was poisoned. Improvements, therefore, required a reduction in the ratio of Pd^{2+} with respect to W^{6+} while maintaining a facile electron transfer between Pd^O and W^{6+}. This was accomplished using a proprietary adaptation to the well-known Wacker process used for reversible chemochromic carbon monoxide sensors. However, reversibility of the IOS hydrogen optrode does not depend on atmospheric oxygen. The result is an optical hydrogen sensor with excellent long-term stability, fast response and a wide dynamic range.

Discussion

Figure 1 shows the general behavior of hydrogen optrodes at hydrogen concentrations ranging from 0.1 percent to 5 percent v/v. Table 1 provides a key to the concentrations. In this test the optrode required five seconds to achieve 90 percent of full response.

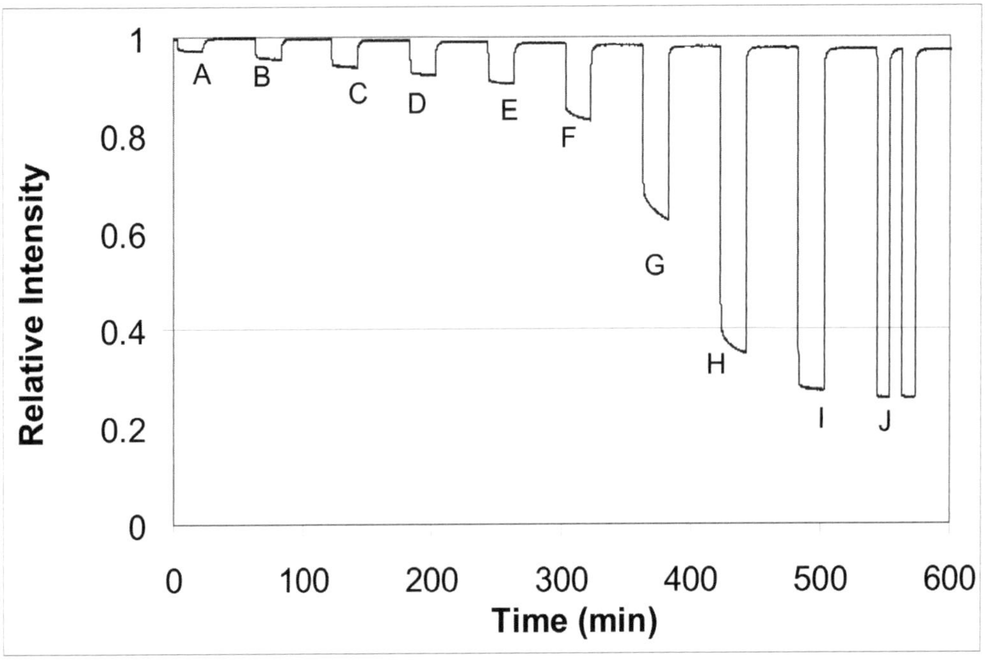

Figure 1. The sensor response as a function of hydrogen concentration. The diluent gas used was Helium. The key to the hydrogen concentrations is shown in Table 1 below.

Table 1	
Figure Key	Percent Hydrogen
A	0.1
B	0.2
C	0.4
D	0.6
E	0.8
F	1.0
G	1.5
H	2.0
I	3.0
J	5.0

These data are revisited in Figure 2, which shows the signal change as a function of hydrogen concentration. This figure shows a roughly linear response in the range of 1 – 3 percent, where the palladium component of the sensing chemistry undergoes the well-known $Pd^{\alpha \rightarrow \beta}$ phase transition. At higher hydrogen concentrations sensor sensitivity falls off asymptotically, while at lower concentrations an "S" shape curve is observed.

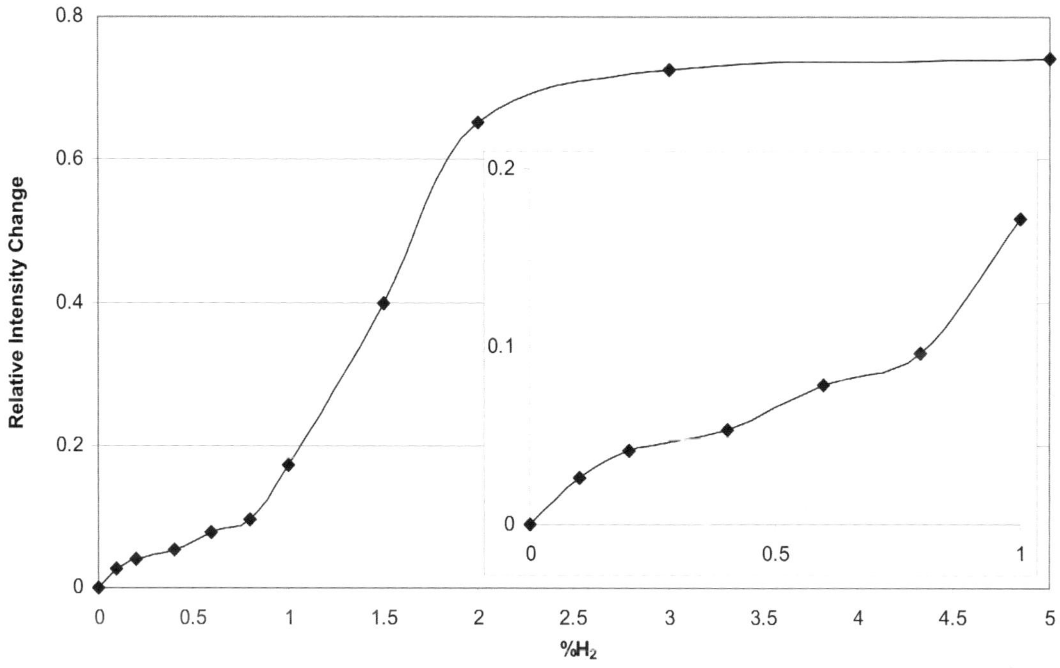

Figure 2. Hydrogen Optrode signal change as a function of hydrogen concentration.

The potential sensitivity of the optrodes is shown in Figure 3. In this experiment a 1.4 percent change in optrode transmissivity was measured in response to 10 ppm H_2 in a helium carrier gas. The signal to noise ratio was about 50:1, and on this basis a detection limit at least as low as 1 ppm is estimated, with an upper range of measurement as high as 5 percent hydrogen in the same sensor.

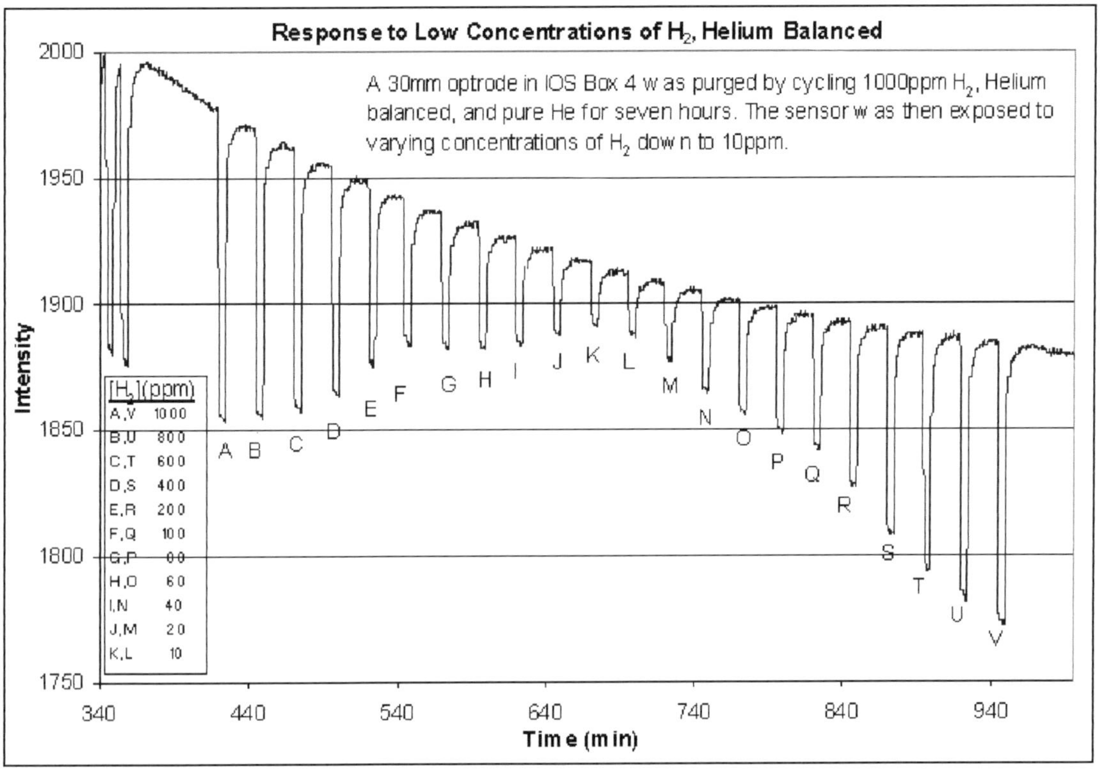

Figure 3. Hydrogen Optrode response to various concentrations of hydrogen down to 10 ppm.

The impact of moisture on the hydrogen optrode signal was evaluated for two reasons. First of all, Vycor is known to be hydrophillic, and adsorption of water in pores is likely to change the optical properties of the matrix. Secondly, some of the process steps in optrode preparation involve aqueous chemistry, and so condensed water has the potential to modify the sensing system. In these tests an optrode was alternately exposed to nitrogen and five percent hydrogen in nitrogen. The humidity level was set at 90 percent at 20 °C. Figure 4 summarizes the key results of this study. It shows that in the presence of high humidity the sensor signal degrades with time. However, even though the overall signal decreases, the calculated absorbance remains approximately constant, as long as the signal-to-noise ratio remains favorable. As a result, water is found neither to interfere with the hydrogen response of the sensor and does not confound the signal. This suggests that in some circumstances humidity can be tolerated, as long as detector software is designed with sufficient robustness to compensate for humidity-induced baseline signal drift.

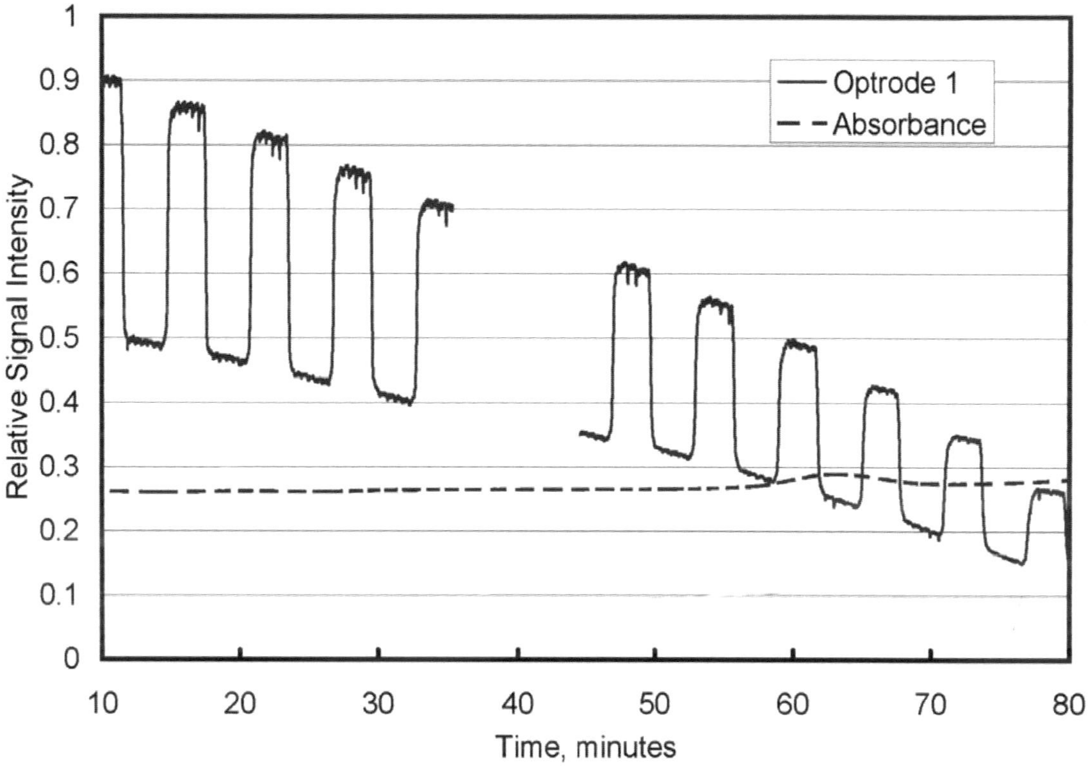

Figure 4. Effect of humidity on response of the hydrogen optrode to 5 percent hydrogen in nitrogen.

Finally, the impact of temperature on hydrogen optrode behavior was investigated. Figure 5 summarizes the data collected in that study. In general it was found that as temperature increased, the upper detection limit of the optrode also increased. At the same time, optrode response time improves. The change in upper detection limit is a direct reflection of gas-solid equilibrium, and is related to the free energy change of the hydrogen-palladium solution formed in the sensor. Changes in response time were found to correlate well with the hydrogen diffusion coefficient. At hydrogen concentrations lower than 1 percent v/v the effect of temperature on signal response can be represented by a simple two-term free energy expression.

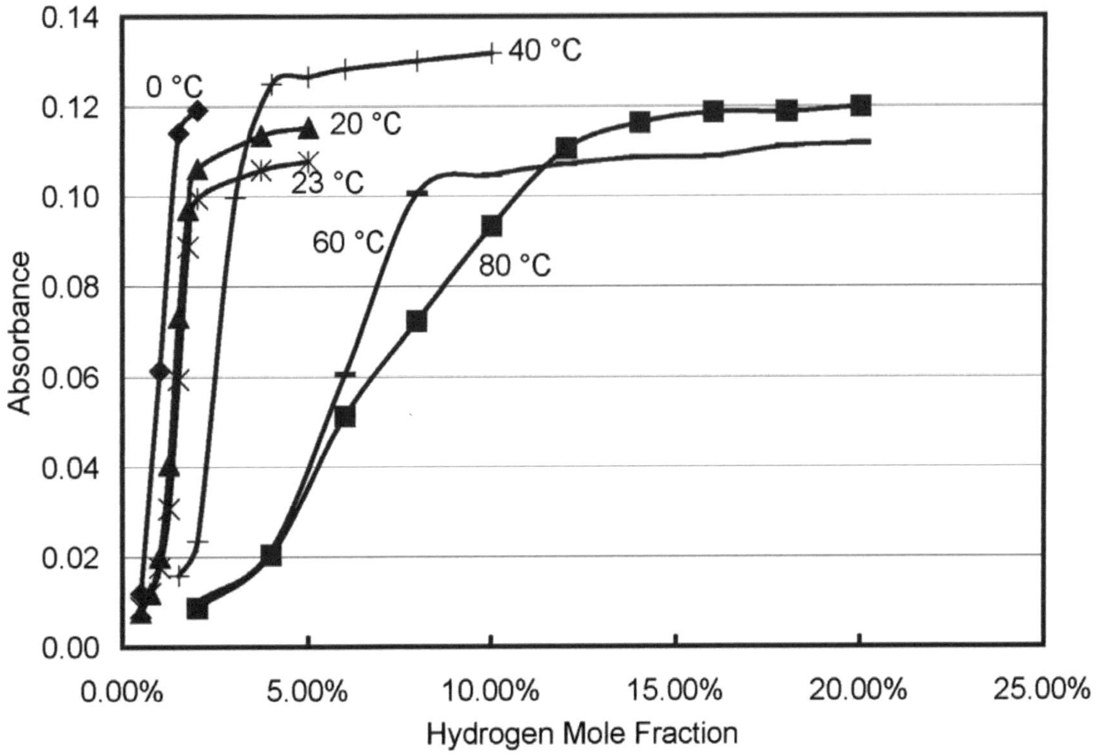

Figure 5. Optrode signal for different temperatures and hydrogen concentrations (x-axis).

Summary

Intelligent Optical Systems has developed a optical hydrogen sensor capable of measuring hydrogen concentrations from as low as 1 ppm to 5 percent at ambient temperature. The sensor operates in inert environments, such as those found in space launch applications, and has a rapid response time on the order of seconds. The sensor also displays a response to temperature that can be readily calibrated and is robust toward ambient moisture.

Acknowledgements

The work described here was supported by NASA contracts NAS10-01065 and NAS10-02039.

References

1. Benson, D. K., C. E. Tracy, G. A. Hishmeh, P. A. Ciszek, Se-Hee Lee, and D. P. Haberman. (1998) "Low-cost, fiber-optic hydrogen gas detector using guided-wave, surface-plasmon resonance in chemochromic thin films," *Proceedings of the SPIE International Symposium on Industrial and Environmental Monitors and Biosensors*, Boston, MA, November 2-5.

2. Smith, R. Davis, Ping Liu, Se-Hee Lee, Ed Tracy and Roland Pitts (2002) "Interfacial Stability of Thin Film Fiber-Optic Hydrogen Sensors" *Proceedings of the 2002 U.S. DOE Hydrogen Program Review NREL/CP-610-32405.*

3. Mendoza, Edgar A. and Anil Menon. (2000) "Hydrogen Sensor Apparatus and Method of Fabrication". U.S. Patent No. 6,535,658.

H2 Detection via Polarography

Jesús A. Dominguez, Ron Barile
ASRC Aerospace Corporation
Kennedy Space Center, Mail Stop ASRC-20.
Kennedy Space Center, Florida 32899

Abstract

Polarography is the measurement of the current that flows in solution as a function of an applied voltage. The actual form of the observed polarographic current depends upon the manner in which the voltage is applied and on the characteristics of the working electrode. The new gas polarographic H_2 sensor shows a current level increment with concentration of the gaseous H_2 similar to those relating to metal ions in liquid electrolytes in well-known polarography. This phenomenon is caused by the fact that the diffusion of the gaseous H_2 through a gas diffusion hole built in the sensor is a rate-determining step in the gaseous-hydrogen sensing mechanism. The diffusion hole artificially limits the diffusion of the gaseous H_2 toward the electrode located at the sensor cavity.

This gas polarographic H_2 sensor is actually an electrochemical-pumping cell since the gaseous H_2 is in fact pumped via the electrochemical driving force generated between the electrodes. Gaseous H_2 enters the diffusion hole and reaches the first electrode (anode) located in the sensor cavity to be transformed into an H^+ ions or protons; H^+ ions pass through the electrolyte and reach the second electrode (cathode) to be reformed to gaseous H_2.

Gas polarographic O_2 sensors are commercially available; a gas polarographic O_2 sensor was used to prove the feasibility of building a new gas polarographic H_2 sensor.

Background.

Gas polarographic O_2 sensors are commercially available; a gas polarographic O_2 sensor was used to prove the feasibility of building a new gas polarographic H_2 sensor. Figure 1 shows a schematic diagram of the commercial gas polarographic O_2 sensor. The sensor consists of an electrochemical O_2-pumping cell and a cap with a cylindrical gas diffusion hole for the artificial diffusion control of the gaseous O_2. A tiny heater is mounted on the cap of the sensor to keep the sensor temperature at 400-450 °C. The O_2-pumping cell has the form of a thick disk of 450-500-μm thickness and 60-70-μm diameter and is made of yttria-stabilized zirconia (YSZ); porous platinum electrodes are attached to both sides of the electrolyte disk. The voltage is applied so that the electrode in the sensor cavity is negative (cathode) to control the diffusion of gaseous O_2 and reduce it to negative O_2 ions. These negative ions pass through the YSZ electrolyte and reach the positive electrode (anode), where they subsequently reform to gaseous O_2.

In existing and commercially available gas polarographic O_2 sensors, the electrode in the sensor cavity has to be negative (cathode) as gaseous O_2 is reduced to negative ions (O^{-2}) at the cathode. In contrast, a gas polarographic H_2 sensor should have the positive electrode (anode) in the sensor cavity as gaseous H_2 needs to release electrons and generate H+ ions or protons. These H^+ ions (or protons) should pass through the YSZ electrolyte to the negative electrode (cathode), where they subsequently reform to gaseous H_2. Figure 2 shows a schematic diagram of the new gas polarographic H_2 sensor. This gas polarographic hydrogen sensor is actually an electrochemical-pumping cell since the gaseous H_2 is in fact pumped via the electrochemical driving force generated between electrodes.

YSZ is used as electrolyte in most of the commercial gas polarographic O_2 sensors as well as in Solid Oxide Fuel Cells (SOFC) because of its excellent oxygen ion (O^{-2}) conductivity. YSZ should be also a protonic (H^+) conductor to use it as electrolyte in the new gas polarographic H_2 sensor. The conduction of protons through solid oxide electrolytes has been subject of some discussion, and it is generally believed to occur through a "hopping" mechanism. In this the conduction occurs by the proton associating with an oxygen atom, forming a transient hydroxyl specie that is "transported" through the material by the migration of the hydrogen ion from one oxygen atom to a neighboring oxygen atom. Since the exact details remain unclear, dynamic quantum mechanical modeling would be needed to understand the mechanism of proton conduction. Recent research has reported electrochemical properties of different ceramic protonic conductors and the feasibility of H_2 pumping for the purpose of tritium extraction in fusion fuel systems.

Sensing Mechanism.

Having a gas binary system and assuming steady-state conditions, no reaction, stagnant background gas, and mass transfer only in one direction and rate-limited at the diffusion hole, the limiting output current is predicted to be a linear function of the logarithm of the background gas concentration. See enclosed appendix A for detailed derivation of this function. This linear relation between the output limiting current and the logarithm of the background molar fraction has a slope that includes the area and the length of the gas diffusion hole, the gas pressure, the gas temperature, the diffusion coefficient of the gaseous specie in the binary gas mixture, the Faraday constant, and the number of electrons per mole of the gaseous specie transferred between electrodes (this number is 4 for O_2 and 2 for H_2 as indicated in figures 1 and 2 respectively).

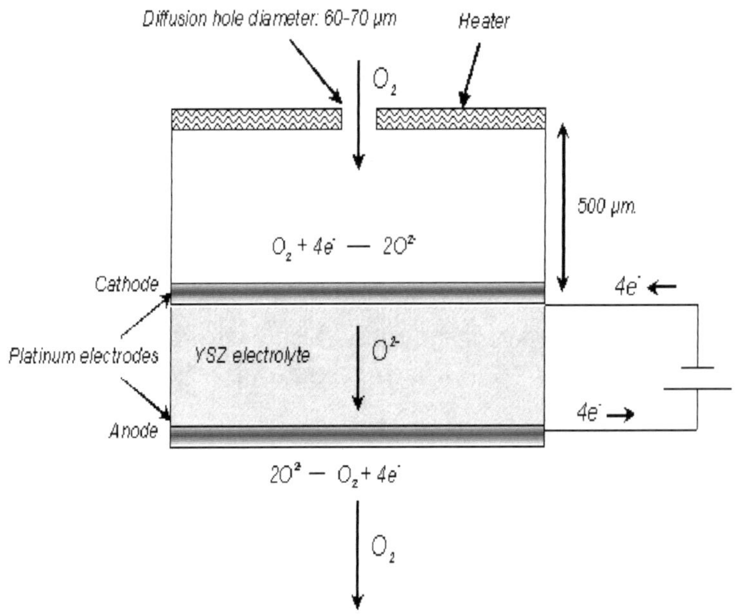

Figure 1. *Schematic diagram of commercially available gas polarographic O_2 sensor.*

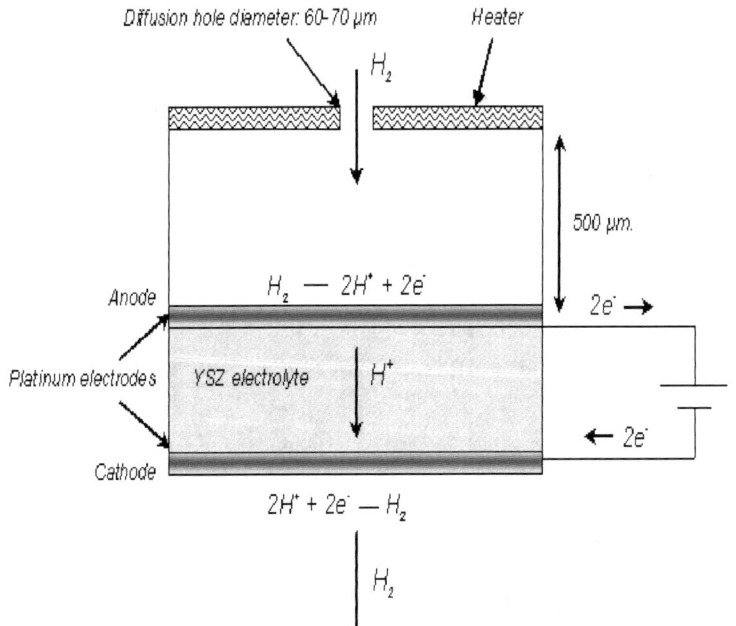

Figure 2. *Schematic diagram of new gas polarographic H_2 sensor.*

It is generally known that YSZ exhibits purely oxygen ionic conduction (with no electronic conduction) over a wide range of oxygen partial pressures. Researchers have corroborated that YSZ is also a protonic conductor. Research work is currently oriented to build a ceramic electrolyte that can perform proton conduction at temperatures lower than those needed with YSZ. Further research might lead to the feasibility of building ceramic materials with hydrogen pumping capabilities for the purpose of tritium extraction in fusion fuel systems.

Results.

Plot 1 displays the voltage-current characteristics of the commercial gas polarographic O_2 sensor. Plot 2 shows the voltage-current characteristics of the gas polarographic H_2 sensor in H_2-N_2 gas mixtures (from 0-3% H_2) when voltage is applied so that the electrode in the sensor cavity is positive (see figure 2) in order to control the diffusion of gaseous H_2. As expected, limiting current proved to be dependent on H_2 composition and increased with increasing H_2 concentration in the ambient gas mixture. However, plot 2 does not show defined limiting-current plateaus similar to those shown in plot 1. In both cases we used the same sensor with a difference only on the electrode polarity. The size of the diffusion hole (60-70-µm diameter) was actually designed by the sensor manufacturer to limit diffusion of O_2-size molecules. A gas polarographic H_2 sensor with a smaller diffusion hole (20-30-µm diameter) should lead to defined limiting-current plateaus.

Plots 3 and 4 show output current generated by the gas polarographic H_2 sensor at different ambient H_2 concentrations (from 0.25-2.0% H_2) using 0.3 and 0.6 volts respectively. As in plot 2, plots 3 and 4 show results generated by a gas polarographic H_2 sensor having the positive electrode (anode) in the sensor cavity (see figure 2). Input H_2 concentration and flow rate was set up and controlled through a mult-channel mixer; pressure and temperature were set up and controlled at through a test chamber. The operating conditions used were 6000 sccm, 768 torr, 27 ºC, nitrogen as background gas, and 0.25-2.0% H_2 concentration range. The setup H_2 concentration was verified using a RGA that takes samples continuosly from the test chamber inlet line; RGA readings (H_2 concentration) are drawn using red dotted lines while output current readings are drawn using violet dotted lines As expected, the output limiting current proved to be dependent on H_2 composition; the output limiting current is increased as the gaseous H_2 concentration increases. As in the case of the gas polarographic O_2 sensing, a linear relation between the output limiting current and the logarithm of the mole fraction of the background gas was also observed for the gas polarographic H_2 sensing as shown in plot 5; further test runs and precise measurements of the diameter and the length of the diffusion hole are required to corroborate this linear relation (see appendix A). The output-current response peaks shown in plots 3 and 4 should be attenuated with a smaller diffusion hole; 20-30 µm might be a suitable diffusion hole diameter for a gas polarographic H_2 sensor. As mentioned above, the size of the diffusion hole (60-70-µm diameter) was actually designed to limit diffusion of O_2-size molecules.

Plot 6 shows output current generated by the above sensor but having the negative electrode (cathode) in the sensor cavity as set up in the gas polarographic O_2 sensor. The output limiting current shows no dependency on H_2 concentration proving that limiting-current mechanism on a H_2 pumping cell can be possible only if the positive electrode (anode) is in the sensor cavity.

The above results show that if the diffusion of gaseous H_2 is artificially controlled and a protonic conductor is used as electrolyte, a gaseous H_2 sensor might be built based on a gaseous H_2-sensing mechanism attained on the analogy of the widely known O_2-sensing mechanism.

Research and development on solid materials that can perform proton conducting at intermediate temperature have dramatically increased due to the renewed interest in the industry for solid-electrolyte fuel cells. Commercially available Proton Exchange Membranes (PEM) might be utilized as solid-electrolyte in the gas polarographic H_2 sensor. PEM performs proton conducting at relatively low temperature (5-120 ºC). Research on new zirconia-based electrolyte materials are currently focus on obtaining materials that can perform proton conducting at intermediate temperature; YSZ, the electrolyte currently used in SOFC and commercial gas polarographic O_2 sensor, requires temperatures higher than 400 ºC. Figure 3 shows the proton conductivity of BCY and SCZT85 (two new zirconia-based materials) along with YSZ proton conductivity[1].

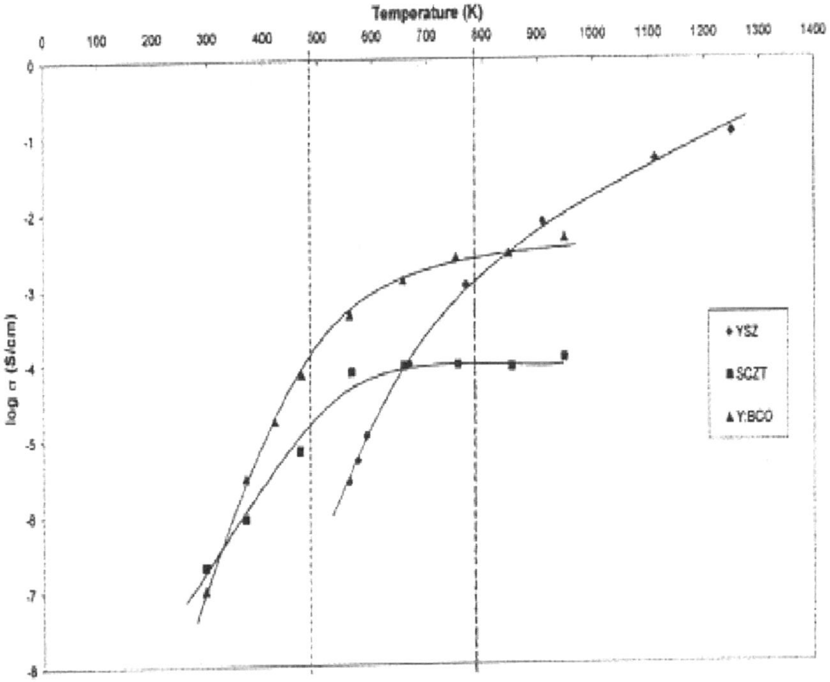

Figure 3. *Proton Conductivity of BCY, SCZT85 and YSZ with Temperature[1].*

Conclusions

The engineering and manufacturing requirements of this new gas polarographic H_2 sensor would be basically the same as those already in place for the commercially available gas polarographic O_2 sensors. The gas polarographic H_2 sensor would need a smaller diffusion hole and the anode in the sensor cavity. Further research is needed to determine the proper size of the diffusion hole and evaluate different types of solid electrolytes including electrolytes with more H+ ion and less O-2 ion conduction selectivity and capable of conducting protons at lower temperatures.

An array of two gas polarographic sensors with different diffusion hole sizes and voltage polarities might be able to detect three components—O_2, H_2, and humidity. A smart sensing system equipped with the proper algorithm might be able to extract mutual cross-sensitivity interference of H_2 and humidity on the O_2 sensor and O_2 and humidity on the H_2 sensor.

Based on the results shown in this H_2 sensing approach along with what is already known from the gas polarographic O_2 sensor, the concept of polarography should be reconsidered and redefined to be classified in two types—gas polarography and the liquid polarography—on the basis of the phase of the measured substance, the diffusion of which is a rate-determining step. For example, the concept of chromatography is well defined and clearly divided into the gas chromatography and liquid chromatography because of the kind of mobile phase. Thus, taking analogous consideration, this proposal is clearly reasonable.

References.

[1] D.Browning, M. Weston, J.B. Lakeman, P. Jones, M. Cherry, J. T. S. Irvine, D. J. D. Corcoran; "Proton Conducting Ceramics for Use in Intermediate Temperature Proton Conducting Fuel Cells", Journal of New Material for Electrochemical Systems, 5, page 28 (2002).

[2] K. Lundstrom, M. Shivaraman, and C. Svenson, J. Apply. Phys., 46, 3976 (1975).

[3] Y. Tan, T. C. Tan; " Sensing Behavior of an Amperometric Hydrogen Sensor", J. Electrochem. Soc., Vol 142, No 6, June 1995.

[4] T. Usui, A. Asada, K. Ishibashi, M. Nakazaka, "Humidity Sensing Characteristics in a wet air of a gas polarographic Oxygen Sensor Using Zirconia Electrolyte", Electroshem. Soc. 138 (1991) 585-588.

[5] J. Gopaul, W. C. Maskell, K. E. Pitt, "Planar Oxygen Sensor: Part I. Effect of Crazing of Zirconia Thick Film on an Alumina Substratem J. Appl. Electrochem., 29 (1999) 93-100.

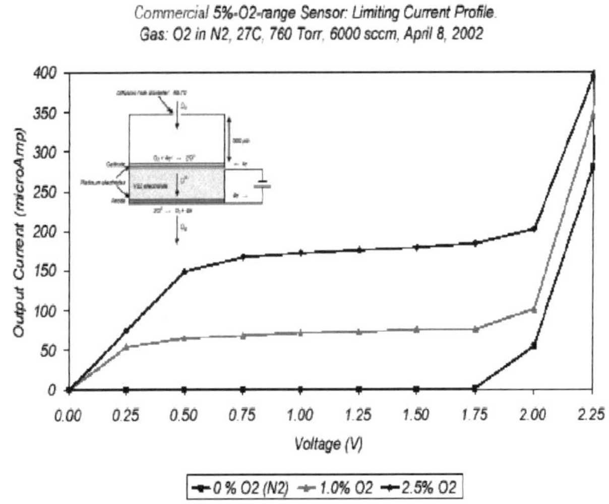

Plot 1. O_2 Limiting Current Profile

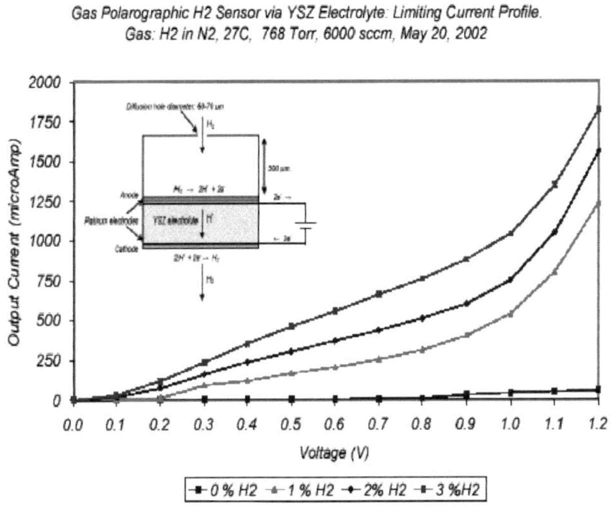

***Plot 2.** H_2 Limiting Current Profile.*

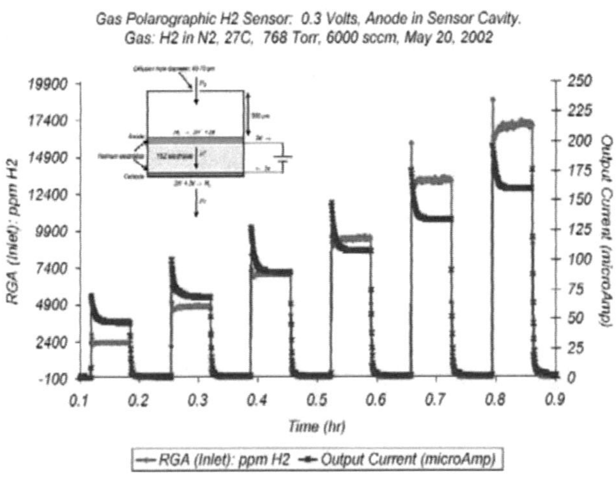

***Plot 3.** H_2 measurement profile at 0.3 V with anode at the sensor cavity.*

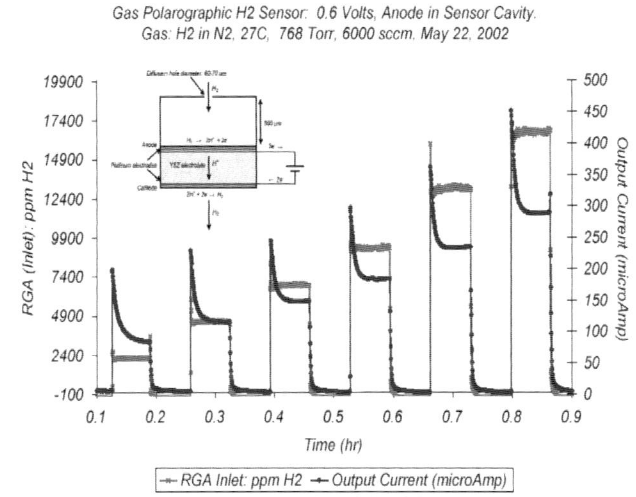

Plot 4. *H_2 measurement profile at 0.6 V with anode at the sensor cavity.*

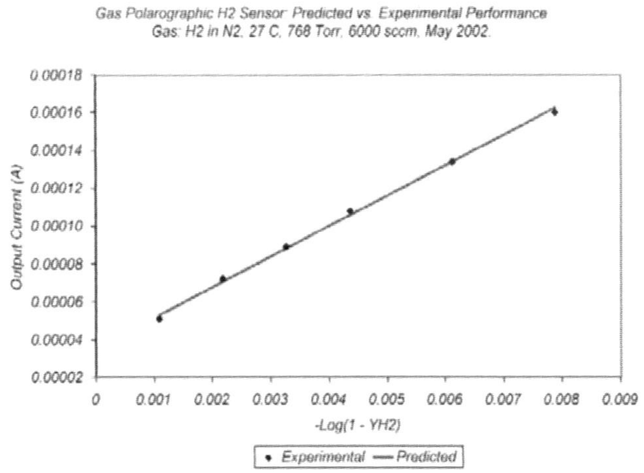

Plot 5. *Experimental vs Predicted results.*

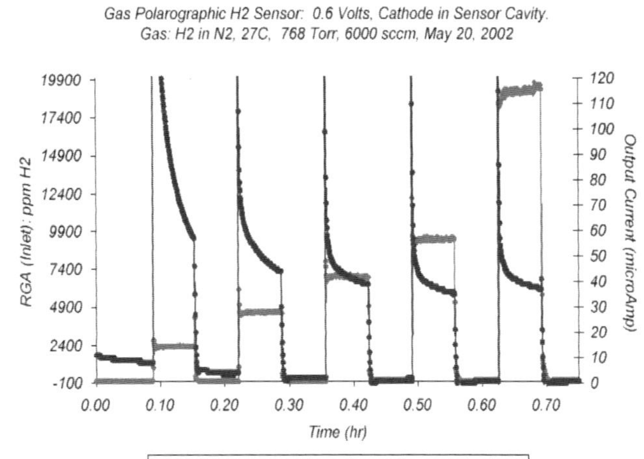

Plot 6. *H_2 measurement profile at 0.3 V with cathode at the sensor cavity.*

APPENDIX A

Gas Polarographic Sensor: Theoretical Mechanism

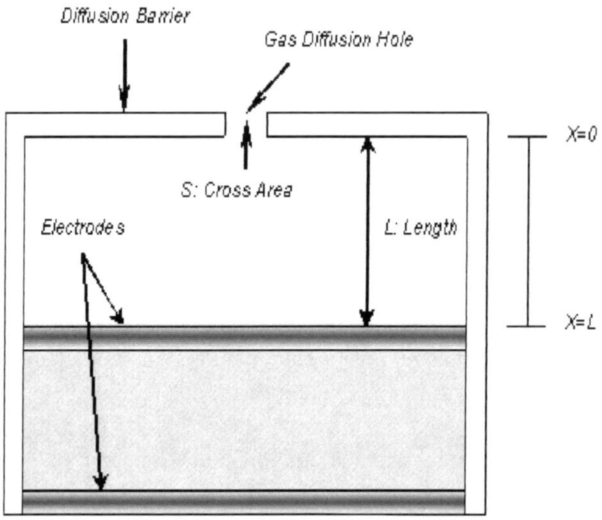

Assuming a binary mixture (A and B where B is the background gas). The limiting current is expressed as:

$$I_l = \eta F S \overbrace{N_A}^{x=L} \qquad (1)$$

Where

I_l	Limiting current.
η	Number of electrons per mole of specie A transferred from the anode to the cathode. η = 4 if A specie is O_2 ($O_2 + 4e^- \rightarrow 2O^{-2}$). η = 2 if A specie is H_2 ($2H^+ + 2e^- \rightarrow H_2$).
F	Faraday constant (electrical charge of a mole of electrons).
S	Cross area of diffusion barrier.
$\overbrace{N_A}^{x=L}$	Molar flux of specie A at the electrode.

The general equation for molar flux of specie A (N_A) can be expressed as:

$$\frac{\partial C_A}{\partial t} + \left(\frac{\partial N_{Ax}}{\partial x} + \frac{\partial N_{Ay}}{\partial x} + \frac{\partial N_{Az}}{\partial x} \right) = R_A \quad (2)$$

Where

C_A	Molar concentration of specie A.
N_{Ax}, N_{Ay}, N_{Az}	Molar flux of specie A in x, y, and z directions respectively.
R_A	Reaction of specie A.

Assuming steady state conditions, no reaction, and mass transfer only in one direction (x direction) equation (2) becomes:

$$\frac{\partial N_{Ax}}{\partial x} = 0 \quad (3)$$

Assuming a binary system denoting A as the component being measured and B as the background gas, N_{Ax} can be expressed as:

$$N_A = -CD_{AB}\frac{\partial Y_A}{\partial x} + Y_A(N_A + N_B) \quad (4)$$

Where:

N_A	Molar flux of specie A.
N_B	Molar flux of specie B.
C	Molar concentration.
D_{AB}	Diffusion coefficient of A into B.
Y_A	Molar fraction of specie A.

Assuming specie B (background gas) is stagnant ($N_B = 0$), equation 4 becomes:

$$N_A = -CD_{AB}\frac{\partial Y_A}{\partial x} + Y_A N_A \quad (5)$$

Rearranging equation (5):

$$N_A = -\frac{CD_{AB}}{(1-Y_A)}\frac{\partial Y_A}{\partial x} \qquad (6)$$

Substituting equation (6) into equation (3):

$$\frac{\partial\left(\frac{CD_{AB}}{(1-Y_A)}\frac{\partial Y_A}{\partial x}\right)}{\partial x} = 0 \qquad (7)$$

Integrating (7) once:

$$\frac{1}{(1-Y_A)}\frac{\partial Y_A}{\partial x} = k_1 \qquad (8)$$

After separating variables and integrating again:

$$Log_N(1-Y_A) = k_1 x + k_2 \qquad (9)$$

Using the boundary conditions $Y_A = Y_{A(amb.)}$ at x=0 and $Y_A = Y_{A(x=L)}$ at x=L we find constant K_1 and K_2. $Y_{A(amb.)}$ is the molar fraction of specie A in the ambient and $Y_{A(x=L)}$ is the molar fraction of specie A at the bottom of the diffusion hole and the top of the electrode.

$$k_1 = \frac{1}{L} Log_N \frac{\left(1-\overbrace{Y_A}^{x=0}\right)}{\left(1-\overbrace{Y_A}^{x=L}\right)} \qquad (10)$$

$$k_2 = -Log_N\left(1-\overbrace{Y_A}^{x=0}\right) \qquad (11)$$

Substituting (10) into (8):

$$\frac{\partial Y_A}{\partial x} = \frac{(1-Y_A)}{L} Log_N \frac{\left(1-\overbrace{Y_A}^{x=0}\right)}{\left(1-\overbrace{Y_A}^{x=L}\right)} \qquad (12)$$

Substituting (12) into (6):

$$N_A = -\frac{CD_{AB}}{L} Log_N \frac{\left(1-\overbrace{Y_A}^{x=0}\right)}{\left(1-\overbrace{Y_A}^{x=L}\right)} \qquad (13)$$

Assuming:

1) $\overbrace{Y_A}^{x=L}$ << 1 << $\overbrace{Y_A}^{x=0}$ (A is consumed at the electrode).

2) $\overbrace{Y_A}^{x=0}$ = Y_A (A concentration in environment).

3) $C = \frac{P}{RT}$ (Ideal gas).

equation (13) becomes:

$$N_A = -\frac{PD_{AB}}{RTL} Log_N (1-Y_A) \qquad (14)$$

Substituting (14) into (1):

$$I_l = -\frac{\eta FSPD_{AB}}{RTL} Log_N (1-Y_A) \qquad (15)$$

Equation (15) states that the limiting current can be used to measure the composition of specie A since there is a linear relation between I_L (limiting current) and $Log_N(1-Y_A)$. The slope of this linear relation is $-\frac{\eta FSPD_{AB}}{RTL}$ leading to a limiting current dependence on the geometric of the diffusion hole (L = length and S = Cross area), pressure P, temperature T, specie A itself

(η = number of electrons per mole of specie A transferred from the anode to the cathode), and D_{AB} = diffusion coefficient of A into B. To avoid the need of using these parameters and assuming that T, P, and D_{AB} remain unchanged during the measurement, equation (15) may be reduced to:

$$I_l = I_l^* \frac{Log_N(1 - Y_A)}{Log_N(1 - Y_A^*)} \qquad (16)$$

Where

I_l	Limiting current at Y_A.
Y_A	Mole fraction of specie A.
I_l^*	Reference limiting current at Y_A^*.
Y_A^*	Reference mole fraction of specie A.

The reference limiting-current value ($I_L^{(Ref.)}$) is often measured using a calibration gas containing a known composition value ($Y_{A(amb.)}{}^{(Ref.)}$) that is equal to the upper composition value of the sensor's range.

Development of Hydrogen Detection Instruments at Kennedy Space Center

Brian Hunter
C Richard Arkin
Charles Curley
David Floyd
Guy Naylor
ASRC Aerospace Corporation

The growing hydrogen economy has ever increasing instrumentation needs. Chief among them is the detection of hydrogen leaks and dangerous conditions. The poster presented discusses the past, present, and future projects of the Hazardous Gas Detection Lab and how they fit into the instrumentation needs of the developing hydrogen economy.

Introduction

Hydrogen may be the perfect fuel for propulsion of spacecraft into orbit. On a pound for pound basis, the combustion of hydrogen releases more energy than any other non-nuclear fuel. This allows the vehicle to spend less of its gross payload capacity on fuel, and more on mission-specific payloads. When in its liquid form, its energy density is even more favorable. Since it is so useful as a fuel, Kennedy Space Center has a large number of hydrogen handling facilities. As oil supplies tighten, the entire world is also realizing the benefits of hydrogen. Development of the hydrogen economy promises to not only provide affordable energy, but pollution-free energy as well.

Unfortunately, hydrogen is not without its risks: it requires very little energy to ignite, is flammable over a wide range of concentrations, and burns with a pale blue, almost invisible, flame. Additionally, hydrogen is known to embrittle and crack certain metals, and because of to its high molecular velocity and small size, hydrogen easily escapes through tiny cracks and even elastomeric seals. Indeed, the risk to personnel and equipment is extremely high.

Techniques and Applications

Since lives and property can be endangered with hydrogen, detecting it is a priority at Kennedy Space Center, and will become a huge priority for anyone involved in the hydrogen economy. Most hydrogen detection techniques fall into two categories, instrument based methods and sensor based methods, for lack of better names. Sensors are small, rugged devices that operate on a simple chemical or physical change in a material. Instrument methods involve complex devices with many subsystems.

Detectors that fall into the "sensor" category can operate on a variety of techniques. Some sensors rely on hydrogen permeation through a thin film of palladium. This permeation changes the properties of the palladium or another material inside of the palladium boundary, which is then measured. Because this technique is diffusion based, it is temperature sensitive and has sluggish response times and long recovery times. Other sensors work on the reaction of hydrogen with oxygen in the air. This redox reaction generates a voltage which is measured. This technique is not specific only to hydrogen, nor can it be used in the purged areas commonly found in aerospace and chemical processing. Other techniques appear simple at first, but end up being as complicated as analytical methods. Color-changing hydrogen sensors are sometimes viewed remotely with a spectrophotometer to quantify concentration, which defeats the purpose of having a small point sensor in the first place. Nevertheless, point sensors are useful. They are light, rugged, inexpensive, and easy to use. Their development continues to show promise.

Conversely, instrument based analytical techniques, mass spectrometry in particular, are heavy, fragile, expensive, and require highly trained operators. What they give up in convenience, however, they gain in capability. Mass spectrometry is the single most specific and accurate analytical technique in existence. With proper sample delivery and pumping systems, mass spectrometry also has good response and recovery times, and the measurement can be made in virtually any mixture.

The question of which technique to utilize is a question of application. If the user desires to detect a hydrogen leak from a fuel cell powered vehicle, a fragile and expensive mass spectrometer is not a good choice of detector. Not only would it add thousands to the cost of the vehicle, it would likely be damaged the first time the vehicle hit a curb. In this application, a rugged point sensor is highly desirable. At Kennedy Space Center, the Hazardous Warning Labs, including the Hazardous Gas Detection Lab, are charged with hydrogen monitoring for all shuttle activities. In a best case scenario, the consequences of inaccurate information from the detector scrub a launch that didn't need to be scrubbed, costing millions of dollars. In a worst case scenario, the system fails to detect a hazardous condition which results in loss of life and spacecraft. The need for extremely accurate, fast, and specific information dictates the use of mass spectrometry for detection of hydrogen leaks in critical areas.

History and Capabilities at Kennedy Space Center

Throughout the space shuttle program, several hazardous gas detection systems have been fielded. The original hazardous gas detection system, PRIME, was first fielded in 1979. It was an ion-pumped, quadrupole mass spectrometer based detector. Because it was ion pumped, it was limited to gas detection in nitrogen or air backgrounds. In December, 1982 during STS-6 flight readiness firing, PRIME detected a 6,000 ppm hydrogen reading, indicating a propellant leak. The test was repeated with the same results and the launch was subsequently scrubbed. PRIME was credited with saving the vehicle.

Having proved itself valuable, a backup HGDS was built to provide redundancy with PRIME. This unit was an ion pumped, magnetic sector mass spectrometer system. Again, because it was ion pumped, it could only operate in air or nitrogen backgrounds. Nevertheless, the backup system was credited with saving the Atlas Centaur AC-60 when it discovered an oxygen leak, which indicated a hazardous mixture with hydrogen.

The need to monitor helium purged areas led to the development of the Turbo Mass Spec in 1988. The replacement of the ion pump with a turbo pump allowed the system to analyze in helium backgrounds. The system was only used to support STS 26 before it was replaced with I-HUMS in 1990.

I-HUMS, the Interim Hydrogen Umbilical Mass Spectrometer was a turbo pumped magnetic sector mass spectrometer based system with improved capabilities over the Turbo Mass Spec. It was capable of detecting 25 ppm with 10% accuracy, with a response time of 30 seconds using its longest sample line, which gave enough notice to make a decision to save lives. It detected leaks at the liquid hydrogen fuel line during STS 35 and STS 38, and is credited with saving both vehicles, 14 lives, and over 2 billion dollars. I-HUMS also supported leak testing for the National Aerospace Plane and the X-33 Linear Aerospike Engine.

I-HUMS was a stopgap measure until a permanent system could be built and fully qualified. HUMS, the Hydrogen Umbilical Mass Spectrometer, was first fielded in 1992 to monitor several helium purged areas. HUMS is capable of detecting 25 ppm with 10% accuracy.

PAMS, the Portable Aft Mass Spectrometer, was designed in 1993. It is a helium leak detector designed to verify leak-tightness of components before hydrogen is ever introduced into the system, saving both time and money while making testing safer. Helium works well for this application because it is a small, fast molecule that easily escapes from seals, just as hydrogen does.

In 2000, obsolescence issues dictated that PRIME and backup HGDS be replaced with a newer, more capable system. The HGDS 2000 was developed.

It is a fully redundant turbo pumped, quadrupole mass spectrometer based system which consists of two separate mass analyzers sharing one sample delivery system. Each system is calibrated and capable of operating independently of the other system. It is capable of operating in nitrogen or helium backgrounds, and is capable of detecting 25 ppm with an accuracy of 10% or better.

In 2003, the Airborne Volcanic Emissions Mass Spectrometer, or AVEMS, debuted. Designed to be a proof of concept for a portable, fully autonomous, rugged, mass spectrometer, AVEMS attempts to manage the shortcomings of mass spectrometry. It is designed primarily to detect part-per-million level traces of pollution in air. CENAT, the Center for High Technology in Costa Rica, and CARTA (Costa Rican Airborne Research and Technology Applications) partnered with the Hazardous Gas Detection Lab to detect emissions from volcanoes in Costa Rica. AVEMS was successfully deployed on small Cessna aircraft; a high altitude research jet, the WB-57; from cars, and even hand-carried into Turrialba volcano.

Future Direction

The Hazardous Gas Detection Lab is interested in continuing to develop small, lightweight, rugged mass spectrometers that could be used in many different applications. One major area where portable mass spectrometers would be useful is in the development of the hydrogen economy. Although point sensors are often preferable for cost and simplicity, mass spectrometry will remain the choice for safety-critical applications in the large scale hydrogen production and storage facilities that will be necessary.

51a Challenges and Benefits of a Move Towards Hydrogen Fuel

Steven G. Chalk, Energy Efficiency and Renewable Energy, U.S. Department of Energy, 1000 Independence Avenue, SW, Washington, DC 20585
Hydrogen addresses the root cause of energy dependence and greenhouse gas emissions. The Hydrogen Fuel Initiative is conducting a comprehensive research program to develop the scientific underpinnings for the hydrogen economy.

Significant challenges will be discussed: • materials with increased hydrogen storage capacity, • lower cost fuel cell materials with improved durability, and • lower cost of delivered hydrogen.

"Well-to-Wheels" analyses that quantify energy use, greenhouse gas emissions and cost of hydrogen delivered will be presented for several pathways.

DOE, with energy companies and automobile manufacturers are learning by conducting system demonstrations under "real world" operating conditions.

Recent policy developments and R&D accomplishments are discussed.

An Overview of the Nuclear Hydrogen Initiative

Carl J. Sink, Jr. and Amy Taylor
United States Department of Energy

Abstract

As part of the Hydrogen Fuel Initiative, proposed by President George W. Bush in 2003, the U.S. Department of Energy (DOE) is developing the technologies and infrastructure to economically produce, store, distribute, and utilize hydrogen, the Nuclear Hydrogen Initiative is developing technologies to provide large amounts of hydrogen without pollution or greenhouse gases. The Nuclear Hydrogen Initiative is a research and development program within the DOE's Office of Nuclear Energy, Science and Technology (NE) that is using a series of successively larger-scale experiments to demonstrate the commercial-scale, economically-feasible production of hydrogen using nuclear energy.

Program Organization

The DOE Hydrogen Program, under the leadership of Office of Energy Efficiency and Renewable Energy (EERE), is working with the DOE Offices of Fossil Energy (FE), Science (SC), and Nuclear Energy, Science and Technology (NE) to meet the President's hydrogen fuel goal. An integrated *Hydrogen Posture Plan* was issued, which outlines the roles of the various organizations as well as the hydrogen research being conducted within the various DOE offices.

The Nuclear Hydrogen Initiative (NHI) funds research and development activities to demonstrate nuclear-based hydrogen producing technologies, to study potential hydrogen production schemes, and to develop deployment alternatives to meet future needs for increased hydrogen consumption. The program collaborates with the Generation IV Nuclear Energy Systems Initiative (Gen IV) to study potential nuclear energy configurations and evaluate deployment scenarios to meet future needs for increased hydrogen consumption. High operating temperatures and improved efficiencies make several Gen IV systems ideal for producing hydrogen, particularly the Very-High-Temperature Reactor (VHTR) because of its passive safety and high temperatures.

The research funded by NHI is conducted primarily at the national laboratories because of its pre-commercial nature; specifically at Argonne National Laboratory (ANL), Idaho National Laboratory (INL), Oak Ridge National Laboratory (ORNL), Sandia National Laboratories (SNL), and Savannah River National Laboratory (SRNL). To the maximum extent possible, DOE also works directly with industry. However, given the current state of the technology, industrial involvement has been limited. Extensive research on high-temperature materials and heat exchangers is being conducted at the University of Nevada, Las Vegas Research Foundation (UNLVRF) in conjunction with other program partners. Additional University Involvement is encouraged through NE's Nuclear Energy Research Initiative (NERI).

Hydrogen Production Methods

Hydrogen can be produced by splitting water into hydrogen and oxygen. However, the feasibility of emission-free, large-scale production of hydrogen from water is as yet unproven. There are four major technologies that can utilize nuclear energy to make hydrogen:

- **Steam Reforming.** Steam reformation of methane currently produces most (about 95%) of the hydrogen produced in the United States. This process is efficient but has the environmental draw back of producing large quantities of carbon dioxide as a by-product. In addition, valuable primary energy sources are consumed in this process thus doing little to reduce the United States' dependence on foreign energy sources. As a result, the NHI is not investigating any steam reforming applications associated with nuclear power.

- **Conventional Electrolysis.** The maximum environmental benefits of electrolysis are realized when a non-emitting technology, such as nuclear energy, is used to produce the electricity. However, there are inherent inefficiencies in producing hydrogen using electricity alone. NHI is not investigating conventional electrolysis applications; however, EERE is funding projects to demonstrate the feasibility and economics of the technology.

- **Thermochemical water-splitting cycles (TC).** These processes offer the potential for high efficiency hydrogen production at large-scale production rates, but the technology is relatively immature. The NHI is dedicating a substantial portion of its funding to develop and demonstrate the feasibility of high-temperature thermochemical water-splitting cycles. The highest priority cycles are the sulfur-based cycles (sulfur-iodine and hybrid sulfur).

- **High-temperature electrolysis (HTE).** HTE, or steam electrolysis, promises higher efficiencies than standard electrolysis, which is employed commercially today. The new high-temperature design involves many technical challenges, including the development of high-temperature materials and membranes. NHI will address these issues and demonstrate the feasibility of the high-temperature electrolysis of steam.

Current Research on Nuclear Methods

Research and development under the NHI focuses on the development of the high-temperature water splitting technologies that can be driven by advanced nuclear systems and on the underlying science supporting these advanced technologies. Investigating and demonstrating these nuclear-based systems will require advances in materials and systems technology to produce hydrogen using thermochemical cycles and high-temperature electrolysis such as high-temperature and corrosive resistant materials development and advanced chemical systems analysis. This program will move through successively larger-scale experiments, namely laboratory-, pilot-, and engineering-scale.

Thermochemical Cycles

The NHI approach for developing high-temperature thermochemical cycles is to develop baseline cycles while also determining the feasibility of alternate cycles. The baseline thermochemical cycles being considered are the sulfur-iodine and hybrid sulfur cycles. Laboratory-scale experiments will be developed for these technologies, as will any additional cycles that are identified and show enough potential to justify that level of investment. The sulfur-iodine and hybrid sulfur cycles are both baseline cycles because they are the most promising mature cycles, however additional alternative cycles are also being researched.

Sulfur-Iodine Cycle

The Department's research and development on the Sulfur-Iodine (S-I) Cycle is being done primarily in collaboration between the French *Commissariat à l'Energie Atomique* (CEA), General Atomics (GA) and Sandia National Laboratories (SNL) under an International Nuclear Energy Research Initiative (I-NERI) agreement. There is close coordination between the project participants in developing the three component reaction sections – the H_2SO_4 decomposition section (SNL), the HI decomposition section (GA), and the Bunsen reaction (CEA). Each participant is designing and constructing their respective section, and is working to integrate them in a Sulfur-Iodine Integrated Laboratory-Scale Experiment. This experiment will be located at General Atomics and is expected to begin operation in late 2007.

Currently, the sulfuric acid boiler and decomposition components in preparation for integrated laboratory-scale experiments are being demonstrated. In 2005 the ambient pressure H_2SO_4 decomposer test bed experiments were completed, the flash distillation concentrator was designed, and high-pressure components for the component reaction tests were developed. At the same time laboratory-scale demonstrations of extractive and reactive distillation options for the HI_x decomposition section were developed, enabling a decision on the best HI distillation process to use in the integrated laboratory-scale experiment. In France, the CEA has completed construction of the Bunsen reactor and supporting equipment and plans to begin testing it before it is integrated with the other two sections in 2007.

High-temperature inorganic membranes are being developed for use in the separation of SO_2 and O_2 from other chemical species in the high-temperature decomposition of H_2SO_4. This separation has the potential to shift the equilibrium of the reaction resulting in a potentially lower reaction temperature or increased process efficiency. This work, at Oak Ridge National Laboratory, leverages existing membrane technology and high-temperature materials expertise.

The S-I cycle requires numerous other chemical separations. Currently, liquid-liquid extraction and distillation are the primary processes being used. To make the S-I process economically feasible, and reduce recycle of water and iodine in the cycle, better separations processes are required. The use of membranes for dewatering process streams is being investigated. Most importantly, the removal of water from a mixture of water, elemental iodine, and hydriodic acid (HI_x) will be studied. Additional studies exploring the use of a membrane to concentrate sulfuric acid will be initiated in fiscal year 2006. Catalysts are also being developed that will be highly-active and stable in the harsh acidic environments and high temperatures encountered in the S-I cycle. These dewatering membrane and catalyst development activities are being pursued at the Idaho National Laboratory.

Hybrid Sulfur Cycle

The Hybrid Sulfur (HyS) thermochemical cycle task addresses the key technology issues involved in the development of a hybrid sulfur hydrogen production system - including the SO_2 - H_2O electrolyzer design, SO_2/O_2 separation, and the unique materials and process issues associated with the acid decomposition section. Savannah River National Laboratory is developing an electrolyzer that, in conjunction with the sulfuric acid decomposition section being developed for the S-I cycle, can be used in a Hybrid Sulfur Integrated Laboratory-Scale Experiment.

Calcium-Bromine Cycle

The potential of the Calcium-Bromine (Ca-Br) cycle for hydrogen production is also being investigated at Argonne National Laboratory (ANL). First, key technical issues must be addressed to determine the viability of calcium-bromine cycles for nuclear hydrogen production. Two areas of research are currently under investigation: examination of cold plasma or electrolytic steps for the hydrogen generation step instead of the iron bromide/oxide reaction beds in the four-stage calcium-bromine process developed by the University of Tokyo (known as the UT-3 process), and investigation of the feasibility of a continuous molten spray reactor approach for the HBr generation step.

Alternative Thermochemical Cycles

The *Nuclear Hydrogen R&D Plan* identifies several thermochemical cycles that have potential for either high efficiencies or operation at lower temperatures, but were not sufficiently investigated or documented to determine their applicability for NHI. Additional thermochemical cycles have been identified as promising by NHI, EERE, or other research efforts. Potentially promising thermochemical cycles that are compatible with nuclear energy systems are being evaluated at ANL to determine whether further research or development is warranted. After the most promising cycles are identified, ANL will coordinate with universities to analyze and evaluate the potential application of those cycles with nuclear heat sources, including flowsheet analysis of selected cycles as an initial screening assessment of alternatives.

High-Temperature Electrolysis

High-temperature electrolysis (HTE) is being developed as one of the baseline processes for nuclear hydrogen production. All HTE activities, led by Idaho National Laboratory, are coordinated with other DOE research activities on solid oxide fuel cell materials and technology development. HTE has the potential for higher overall process efficiency than conventional electrolysis. The current focus of HTE R&D is on providing a technical basis for estimating efficiency improvements and on the development of an HTE design to provide the basis for cost projections.

Computational Fluid Dynamics (CFD) analyses of individual flow channels in a solid oxide electrolyzer cell (conducted at ANL) are being used to model temperature, current density, and local hydrogen production. In 2004, the integrated performance of an HTE plant and the thermal optimization of the plant through various component arrangements were modeled. In addition, solid oxide electrolyzer cell electrode materials with improved durability in the variable environment of the electrolyzer cell are currently being developed. In February of 2006, Idaho National Laboratory (INL) researchers successfully reached 1,000 hours of production of at least 100 liters of hydrogen per hour--a programmatic milestone in producing hydrogen from water using high-temperature electrolysis.

The program is investigating key technical issues for the development of HTE for nuclear application and developing plans for experimental demonstration at the various scaling levels. Components for laboratory-scale experiments have been fabricated and button cell and stack experiments conducted to evaluate candidate electrolyzer characteristics and performance.

Building on previous work, a high-temperature inorganic membrane for the separation of hydrogen from steam at the outlet conditions of the solid oxide electrolyzer cells is being developed to improve the overall thermal optimization of the plant. Current work includes the use of existing experimental data and steam hydrogen separations at temperatures and pressures that will permit extrapolation to HTE operating conditions.

Systems Interface and Support Systems

The NHI is developing hydrogen production technologies to be coupled with an advanced nuclear energy system being developed in parallel as part of the Generation IV Nuclear Energy Systems Initiative. The key research currently underway in this area is the analysis of the S-I and HTE heat exchanger requirements and analysis of the heat transfer medium. Analysis efforts in both interface areas are coordinated extensively with the efforts conducted by universities, national laboratories (particularly Idaho National Laboratory), industry, and other Federal programs.

A partnership between universities, private industry and national laboratories has been established to identify and test high temperature materials and designs for NHI heat exchangers and other system interface components. The work performed by this partnership will identify candidate materials, and perform corrosion and physical property tests necessary for NHI heat exchangers and process equipment needed for thermochemical cycles and high temperature electrolysis.

Energy Policy Act of 2005

The Energy Policy Act of 2005 (Public Law 109-58) was passed by the United States Congress on July 29, 2005 and signed into law on August 8, 2005. A number of provisions that affect the Nuclear Hydrogen Initiative are summarized below:

- Section 634 provides for the establishment of 2 projects to demonstrate the commercial production of hydrogen at existing nuclear power plants, once it has been determine whether the use of existing nuclear power plants is a cost-effective means of producing hydrogen.

- Section 641 establishes a ''Next Generation Nuclear Plant Project'' that will consist of the research, development, design, construction, and operation of a prototype plant, including a nuclear reactor based on research and development activities supported by the Generation IV Nuclear Energy Systems Initiative to generate electricity, hydrogen, or both.

- Section 645 requires that the high-temperature hydrogen production technology that will be demonstrated with an advanced nuclear reactor be selected by the end of fiscal year 2011.

- Section 805 requires the DOE, in consultation with other Federal agencies and the private sector, to conduct a research and development program on technologies relating to the production, purification, distribution, storage, and use of hydrogen energy, fuel cells, and related infrastructure. DOE, in partnership with the private sector, is directed to conduct programs to address production of hydrogen from diverse energy sources, including: fossil fuels, which may include carbon capture and sequestration; hydrogen-carrier fuels (including ethanol and methanol); renewable energy resources, including biomass; and nuclear energy.

- Sections 812 and 934 require solar energy technologies research to coordinate with activities sponsored by the DOE's Office of Nuclear Energy, Science, and Technology on high-temperature materials, thermochemical cycles, and economic issues related to solar energy.

- Section 952 authorizes research to examine designs for high-temperature reactors capable of producing large-scale quantities of hydrogen.

Path Forward

The NHI has defined an aggressive path to demonstrate hydrogen production from nuclear energy by end of the next decade. The technical challenges to achieving this goal are significant, but the development of emission-free hydrogen production technologies is essential to the long-term viability of a hydrogen economy. Nuclear energy has the potential to play a major role in assuring a secure and environmentally sound source of transportation fuels. The fundamental challenge is to focus finite research resources on those processes which have the highest probability of producing hydrogen at costs that are competitive with gasoline. Both thermochemical and high-temperature electrolysis methods have the potential to achieve this objective, and the small-scale experimental precursors to the integrated laboratory-scale experiments have operated successfully to date and continue to show promise for larger-scale systems.

Initially, a broader research effort involving laboratory-scale demonstrations and analytical evaluations is needed to provide a more consistent and complete assessment on which to base future R&D funding and scaling decisions. Integrated laboratory-scale experiments confirm technical viability, and provide information needed for pilot plant decisions. Pilot plant demonstrations of the selected processes confirm engineering viability and establish a basis for process costs. The NHI program will perform consistent and independent analyses of performance and costs to support the comparative assessments required for technology selection and scaling decisions, and establish effective interfaces with industry and international partners. The development of a portfolio of hydrogen production technologies, including nuclear energy technologies, is vital to strengthen the United States' energy, economic, and national security.

51c Thermochemical Cycles for the Production of Hydrogen from Water Using Nuclear Energy

Robert J. Evans, Electric & Hydrogen Technologies & Systems, National Renewable Energy Laboratory, Golden, CO 804021

Thermochemical reaction cycles powered by heat derived from nuclear energy have the potential for production of a significant fraction of the fuel required for the hydrogen economy. This paper will review the state of the science and engineering and the technical barriers that are being addressed by ongoing work sponsored by the Department of Energy. The main focus of the current DOE sponsored work is on the sulfur iodine cycle, but the hybrid sulfur process and the calcium bromine cycle are also being evaluated for potential scale up in the 2012 time frame. Interest continues in identifying alternative cycles as well that may have some advantages by use of less corrosive chemicals, simpler reaction engineering, higher efficiencies, and ultimately lower cost of hydrogen production. The status of these efforts and future scale up plans and challenges will be discussed.

51d "Sustainability Issues Related to the 'Hydrogen Economy'."

Earl R. Beaver, Practical Sustainability, LLC, PO Box 599, Chesterfield, MO 63006-0599

The Sustainable Engineering Forum of AIChE's Institute for Sustainability uses a Virtual Meeting concept for vetting issues related to Sustainability. Two Virtual Meetings were held in 2005 on "Sustainability issues related to the 'Hydrogen Economy'." More than forty scientists and engineers reviewed current literature on the subject and discussed whether technology advances had obviated many concerns about sources of hydrogen and other issues regarding safety and economics. We also examined whether public and political pressures were shifting to favor hydrogen.

Major questions covered were: What is the driving force for the Hydrogen Economy?; What technologies will be needed to make the switch?; What are the unique issues of "Hydrogen" and how do they differ by region?; How will a hydrogen economy affect national and global infrastructure?; Does the switch make sense?; How can we tell if the switch is the right thing to do?; What sustainability decision tools are needed to choose the correct steps for society as a whole?

This paper will present the salient points developed by this international group of engineers and scientists.

99a Chemical Industry Vision 2020 Overview

Francis Zia, Fairfield Resources Internationnal, 1055 Washington Boulevard, Stamford, CT 06901
A Chemical Industry Vision 2020 sub-committee was established in June 2004 and chartered with the responsibility of identifying research and development needs for separation technology to drive equilibrium processes for the Chemical and Petrochemical Industry. This study focused on the anticipated drivers and requirements of the chemical and petrochemical industries. Chemical production processes were selected for review. This report, the first in a series, is directed to commercial, industrial hydrogen production. The growing requirements for hydrogen in chemical manufacturing, petroleum refining and the new emerging clean energy concepts will place greater demands on sourcing, production capacity and supplies. An overview of the Chemical Industry Vision 2020 will be given during this presentation.

99b Overview of Hydrogen Production Technologies

James A. Ritter and Armin D. Ebner. Department of Chemical Engineering, University of South Carolina, Columbia, SC 29208
A Chemical Industry Vision 2020 sub-committee was established in June 2004 and chartered with the responsibility of identifying research and development needs for separation technology to drive equilibrium processes for the Chemical and Petrochemical Industry. This study focused on the anticipated drivers and requirements of the chemical and petrochemical industries. Chemical production processes were selected for review. This report, the first in a series, is directed to commercial, industrial hydrogen production. The growing requirements for hydrogen in chemical manufacturing, petroleum refining and the new emerging clean energy concepts will place greater demands on sourcing, production capacity and supplies.

A review was conducted of the current state of the art and emerging literature concepts on both adsorption and membrane separation technology applicable to H2 production. Recommendations for future R&D needs are discussed. An emphasis is placed on flow sheet design modification with adsorption or membrane units being added to existing plants for near term impact, and on new designs with complete flow sheet modification for new adsorption or membrane reactor/separators replacing current reactor and separator units in an existing plant for a longer term sustainable impact.

About 41 MM tons/yr of H2 is produced worldwide, with 80% of it being produced from natural gas by steam reforming, partial oxidation or autothermal reforming. H2 is used commercially to produce CO, syngas, ammonia, methanol and higher alcohols, urea and hydrochloric acid. It is also used in Fischer Tropsch reactions, as a reducing agent (metallurgy), and to upgrade petroleum products and oils (hydrogenation). Current industrial practices are summarized in terms of the key reforming and shift reactions and reactor conditions, with the four most widely used separations techniques --absorption, adsorption, membrane and cryogenic. As all of these reactions are reversible, the H2 or CO or syngas production from natural gas and hydrocarbon feedstocks is equilibrium limited and consequently provides opportunities for process improvement

It has been estimated that the reforming of natural gas to produce H2 consumes about 31,800 Btu/lb of H2 produced. It is further estimated that 450 trillion Btu/yr could be saved with a 20% improvement in just the H2 separation and purification train after the H2 reformer. This energy savings is a major driver for future R&D and for implementation of adsorption or membrane separation technology.

Recommendations for future separations R&D are set forth in – Adsorbent Materials Development, Membrane development, Plant Design modification with additional separations systems, and Plant systems optimization. Detailed performance requirements are provided in the report. An overview of this report will be given during this presentation.

Purifying hydrogen with inorganic silica membranes at high temperatures

M. C. Duke, S. Gopalakrishnan, M. Abdel-jawad, M. Macrossan, J. C. Diniz da Costa*

ARC Centre for Functional Nanomaterials, School of Engineering, The University of Queensland, Brisbane Qld 4072, Australia.
**also Project Leader, Centre for Low Emission Technology (cLET), PO Box 883, Kenmore, Qld 4069, Australia*

Abstract: Development of high quality membranes for industrial applications will lead to cost reductions over traditional separations processes. Silica membranes are a new technology for hydrogen separation that needs R&D specifically to applying them to industrial scales. Past work has shown a carbonised template silica membrane which offered hydrostability. This resulted in better stability under steam and high temperature conditions without compromising the permselectivity for small molecules. In this paper a hydrostable silica membrane was developed for hydrogen separation having a pore cut-off around 3Å. The carbon templates did not compromise the membrane's ability to selectively permeate hydrogen over other major gases coming from a synthesised coal gasifier gas mixture of CO, CO_2 and N_2. The selectivity of H_2 to N_2 was 26, whilst the hydrostable property of the carbonised template membrane was maintained. Computational fluid dynamics (CFD) can be used to develop membrane systems in tandem with these intrinsic improvements. CFD simulation studies were also conducted to gain better insight into the macroscopic flow parameters.

Introduction:

Membrane separation of gas mixtures has always considered as a more energy conservative separation method than the conventional separation processes. Hydrogen permselective silica membranes have attracted preference in the membrane gas separation applications due to the importance of hydrogen as a fuel and an industrial feedstock for the production of various chemicals [1]; especially to derive high purity hydrogen for fuel cell applications [2]. Membrane reactors, which combine separations and reactions in one system, benefit from such membranes which shift the reaction conversion limited by thermodynamic equilibrium to the product side, such as processing syngas in the water gas shift reaction [3]:

$$CO + H_2O \leftrightarrow CO_2 + H_2 \qquad \text{(Eq. 1)}$$

By removing H_2 from the reaction system shifts the reaction to the product side deriving increased CO conversion than a conventional catalytic reactor, producing more H_2 while also lowering the operating temperature. This reaction is significant in coal gasification processes which produce syngas, offering significant advantages to cost and size reduction of the operation.

Silica membranes are proven to be thermally and chemically stable offering high permeability and selectivity for hydrogen [4] with much work concentrating on their hydrogen separation property [5]. In silica membranes, the separation is mainly based on a sieving effect through small pores (~3Å) producing reasonably high combined values for flux and selectivity. Based on their mode of formation, silica membranes are mainly divided into sol-gel and chemical vapour deposition (CVD) membranes. Sol-gel membranes are prepared by coating an alkoxide precursor sol atop a porous substrate where the membrane quality is mainly determined by the sol chemistry. CVD membranes are prepared by the decomposition of gas phase precursors inside the micropores of the substrate, thereby blocking the pore diameter to molecular dimensions. The CVD method results in membranes with lower permeance and higher selectivity. However, the main concern regarding the silica membranes is that they are unstable under hydrothermal conditions. A 50% reduction in H_2 permeance under lean hydrothermal conditions (50% H_2O + 50% N_2; atmospheric pressure) was displayed by silica-sol membranes at 500°C [6]. This paper reports a novel concept in which carbon molecules were introduced into the system which position themselves between the silica molecules and act as supporting pillars, thus inhibiting the reorganization of silica molecules under hydrothermal conditions. Another key issue for membrane development is the engineering of separator modules. It is of interest to understand the resulting macroscopic flow parameters such as velocity and pressure distributions for tubes composed from this membrane. Computational fluid dynamics (CFD) simulation studies were conducted to gain insight into the macroscopic flow parameters.

Experimental

Asymmetric supported carbonised template molecular sieve silica (CTMSS) membranes were prepared over porous α-alumina substrates (99.8% purity; porosity = 30%; 0.5 -1.0 µm) by sol-gel dip coating method as described elsewhere [7]. A γ-alumina layer forms the primary layer that brings down the pore size to 4nm, followed by an intermediate layer prepared by dipping a diluted acid catalysed sol-gel solution having methyl triethoxysilane (MTES, Aldrich) precursor leading to a silica with pore sizes in the region 0.35 – 0.45 nm [8]. The final silica layer was prepared by adding a short chained cationic surfactant hexyl triethyl ammonium bromide (Aldrich) to 0.125M of silica sol solution followed by the calcination under vacuum at 500°C, to preserve the hydrophobic organic pyrolysis products. Samples were characterized using adsorption studies and SEM and further tested for gas permeation between 50–200°C (2 atm drop). Permeation was measured using dead-end method to evaluate the performance of single gases involved in Reaction 1 (H_2, CO_2, CO and N_2).

CFD calculations were performed using the commercial code CFD-ACE (ESI Software). The code is a compressible finite volume flow solver. Full Navier Stokes solutions were implemented. Calorically perfect gas was used and viscosity was modelled using kinetic theory. The simulations were axisymmetric. Fixed pressure inflow and outlet conditions were prescribed. Isothermal non-catalytic walls were prescribed at temperatures of 100°C. No chemical reactions were modelled for this set of simulations. Structured grids were used for this simple axisymmetric geometry and these were created using the commercial package CFD-GEOM (ESI Software) which is typically used in conjunction with CFD-ACE. Computational grids for the tubes were generated with clustering near the walls. Grid numbers were minimized as computational resources were limited. The solution was run on a single Pentium IV 3.4 GHz processor. Simulations typically required 16 hours each. The inner tube diameter was 6mm.

Results and discussion

CTMSS membrane permeation through a planar substrate is shown in Figure 1. In general, permeation increases with temperature for all gases except for CO_2. This trend exhibits the classical activated transport mechanism whereby diffusion occurs through micropores, which is in accordance with other high quality silica membranes as reported by other research groups[9, 10]. The CO_2 permeation was essentially unrelated to temperature and decreased with temperature as observed elsewhere [8, 11]. CO_2 has a high heat of adsorption, allowing for a more adsorption based transport. However as the pores become smaller in high quality membranes, CO_2 transport is mostly restricted by the pore size and will also be activated [12]. In this result, no activation with temperature was observed due to two competing mechanisms. The separation mechanism in these tight pore spaces is by molecular sieving.

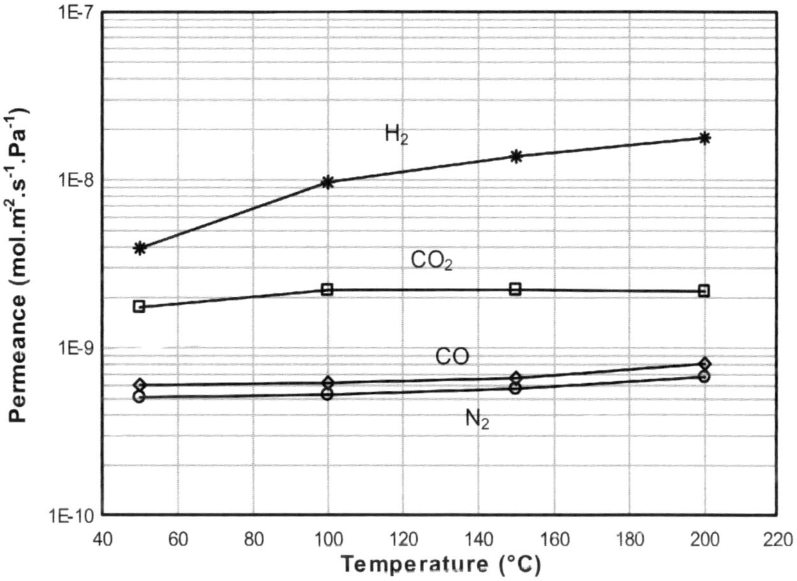

Figure 1: Permeation trend with temperature for planar CTMSS substrate at 2 bar pressure drop.

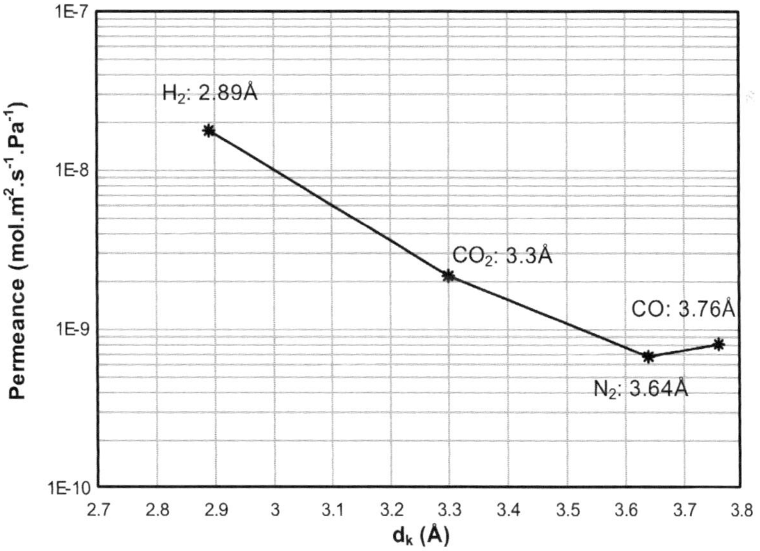

Figure 2: Permeation-molecule size cutoff curve for CTMSS membrane at 200°C, 2 bar pressure drop.

The kinetic diameter of the molecule pertinent to molecular scale diffusion can be plotted against the experimentally measured permeation as shown in Figure 2. The smaller, more mobile H_2 molecule was more permeable than the larger CO_2, CO and N_2 molecules. As most of the permeation dropped between H_2 (2.89Å) and CO_2 (3.3Å), the average pore size is around that of H_2 (~3Å). The permeation of N_2 was lower than CO contrary to the size relationship. Since the permeation of both these gases occurs in the tail of the silica pore size distribution, there is little size selective advantage available. Since CO is a more adsorbing gas than N_2, its adsorptive property therefore favoured its transport through the membrane. The H_2/N_2 permselectivity of the membrane was 26, well above the ideal Knudsen value of 3.74, is another indicator of the molecular sieving behaviour. CVD membranes has however been reported to have far greater selectivity values >1000[13,14].

Figure 3 shows the CFD simulation of the non-dimensionalised velocity field and the centre line non-dimensional pressure and velocity curves (bounded by the 2 crosses in the figure). In this simulation the feed side at the top of the figure has a prescribed velocity. Flow passes from the feed side to the permeate side and exits the permeate side at the far right of the permeate region. The figure shows that the boundary layer size is significant and extends the full radius of the tube and that the velocity profile in the tube axially is linear along the centre line of the flow. This is an indication that the model is scalable in length provided that the flow remains subsonic, but that scaling radially would be more effective. Associated with this is the parabolic pressure distribution dropping from the maximum stagnation pressure at the wall to the left of the figure.

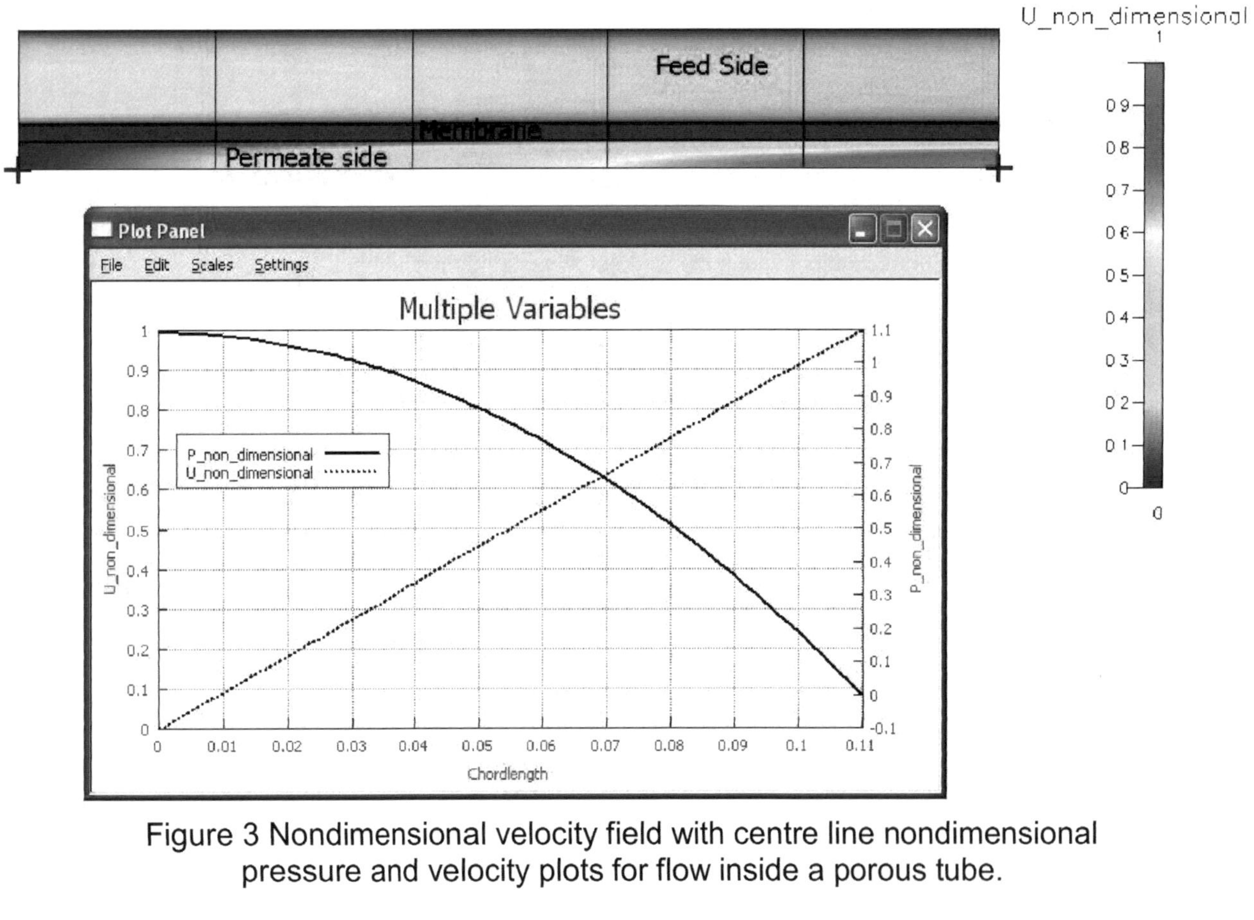

Figure 3 Nondimensional velocity field with centre line nondimensional pressure and velocity plots for flow inside a porous tube.

Conclusion

CTMSS membranes were prepared over asymmetric α-alumina substrates coated with γ-alumina layer. Except CO_2, all other gases studied displayed activated transport through this membrane, indicating molecular sieving effect through this membrane. H_2/N_2

permselectivity of this membrane was 26 and has in past work been found to be quite stable under hydrothermal conditions. CFD studies revealed that the boundary layer size is quite significant and extends the full radius of the tube, and that the velocity profile in the tube axially is linear along the centre line of the flow.

Acknowledgement

The authors acknowledge financial support by the Centre for Low Emission Technology (www.clet.net) in Australia.

References

1. Ramachandran r., Menon R.K. An overview of industrial uses of hydrogen, Int. J. Hydrogen Energy 23 (1998) 593.
2. Ahmed S., krumplet M. Hydrogen from hydrocarbon fuels for fuel cells, Int. J., Hydrogen Energy 26 (2001) 291-301.
3. Kikuchi E. Membrane reactor application to hydrogen production, Catal Today 56 (2000) 97-101.
4. Burggraaf A.J., Cot L., Fundamentals of inorganic membrane science and technology, Elsevier, Amsterdam 1996.
5. Lee D., Zhang L., Oyama S.T., Niu S., Saraf R.F. Synthesis, characterization and gas permeation properties of hydrogen permeable silica membrane supported on porous alumina, J. Membr. Sci. 231 (2004) 117-126.
6. Asaeda M., Kashimoto M. Sol-Gel Derived Silica Membranes for Separation of Hydrogen at High Temperature -Separation Performance and Stability against Steam, Proceedings of the International Conference on Inorganic Membranes, Nagoya, Japan 22-26 June, 1998, 172-175.
7. Duke M.C., Diniz da Costa J.C., Lu G.Q., Petch M., Gray P. Carbonized template molecular sieve silica membranes in fuel processing systems: permeation, hydrostability and regeneration, J. Membr. Sci., 241 (2004) 325-333.
8. de Vos R.M., Maier W.F., Verweij H. Hydrophobic silica membranes for gas separation, J. Membr. Sci. 158 (1999) 277-288.
9. de Lange, R. S. A., Hekkink, J. H. A., Keizer, K. & Burggraaf, A. J. 1995, 'Permeation and separation studies on microporous sol-gel modified ceramic membranes', *Microporous Materials*, vol. 4, no. 2-3, pp. 169-186.
10. de Vos, R. M. & Verweij, H. 1998, 'Improved performance of silica membranes for gas separation', *Journal of Membrane Science*, vol. 143, no. 1-2, pp. 37-51.
11. Duke, M. C., Diniz da Costa, J. C., Lu, G. Q., Petch, M. & Gray, P. 2004, 'Carbonised template molecular sieve silica membranes in fuel processing systems: permeation, hydrostability and regeneration', *J. Memb. Sc.*, vol. 241, no. 2, pp. 325-333.
12. Diniz da Costa, J. C., Lu, G. Q., Rudolph, V. and Lin, Y.S., 'Novel molecular sieve silica (MSS) membranes: characterisation and permeation of single-step and two-step sol-gel membranes', *J. Memb. Sc.*, 198, 9-21, 2002
13. Suraj G., Ono K., Nomura M., Sugawara T., Nakao S. Microporous silica membrane prepared using TMOS/O3 CVD in opposing reactants geometry, Trans. Mat. Res. Soc. Japan 29 (2004) 3263-3266.
14. Suraj. G., Nomura M., Sugawara T., Nakao S. Preparation of multi-membrane module for high temperature hydrogen separation, Desalination (in press).

99d Fast Cycle Rotary Adsorption Bulk Separator at Low Pressure

Surajit Roy[1], Bowie Keefer[2], Aaron Pelman[3], Andre Boulet[4], Matt Babicki[4], and Graeme Herring[4]. (1) QuestAir Technologies Inc., 6961 Russell Avenue, Burnaby, BC V5J4R8, Canada, (2) Advanced Development, QuestAir Technologies Inc., 6961 Russell Avenue, Burnaby, BC V5J4R8, Canada, (3) Process Engineering, QuestAir Technologies Inc., 6961 Russell Avenue, Burnaby, BC V5J4R8, Canada, (4) ADG, QuestAir Technologies Inc., 6961 Russell Avenue, Burnaby, BC V5J4R8, Canada

The separation of gaseous mixtures at low pressures into their constituent components has been a challenge as it is energy intensive and costly. The paper addresses the technology of rotary adsorption to perform separations at low pressures, utilizing the partial pressure difference and a displacement purge (may also include inert purge) stream. This may be applied to hydrogen enrichment from syngas mixtures, where carbon dioxide is to be rejected with nitrogen-enriched air as the displacement purge stream containing residual oxygen/nitrogen. This technology has diverse applications like High Temperature Fuel Cell (HTFC, such as Molten Carbonate Fuel Cell and Solid Oxide Fuel Cell) off-gas, PSA off-gas, CO2 removal and other bulk separations.

Examples will be discussed such as the application to HTFC anode exhaust: · Enables bulk separation of CO2 from the HTFC anode exhaust stream and overall improved hydrogen recovery efficiency. · Utilizes external purging instead pressure-swing for adsorbent regeneration o Bulk separator H2 recovery of over 90% is achievable o Operates at near atmospheric pressure with low pressure drop

Benefits of Bulk Separation Rotary Adsorption Module will be examined: · Enrichment the H2 concentration in HTFC anode exhaust · Decreases in the down stream H2 purification PSA operating pressure · Bulk separation operation at < 3 psig and requiring minimal operating energy · Fast-cycle structured adsorbent bed and rotary valve technology increases the bulk separator's hydrogen production per unit volume and enables a small, efficient device

Hydrogen production via Pd-Ag permeators: membrane reactor design and process study

S. Tosti[1], A. Basile[2], L. Bettinali[1], F. Borgognoni[3], C. Rizzello[3]

[1] ENEA, Unità Tecnico Scientifica Fusione, C.R. ENEA Frascati, Via E. Fermi 45, Frascati, Roma I-00044, Italy, Phone: +39 06 94005160 – Fax: +39 06 94005374, tosti@frascati.enea.it

[2] Institute on Membrane Technology, ITM-CNR, c/o University of Calabria, Via P. Bucci, Cubo 17/C, I-87030 Rende (CS) Italy. Phone: (+39) 0984 492013 – Fax: (+39) 0984 402103.

[3] Tesi Sas, Via Bolzano 28, Roma, Italy Phone/Fax: (+39) 0697273277

ABSTRACT

Thin wall Pd-Ag permeator tubes have been produced by a diffusion welding procedure: these low-cost membranes are proposed for separating and producing high-pure hydrogen in membrane reactors. The reliability of these dense metallic permeators is strongly related to the design configuration of the membrane modules. In fact, as a consequence of hydrogen and thermal cycling, the dense metallic tubes vary their length: therefore, in case of constrains between the membrane and the module, cyclic axial stresses on the tube can rise and involve the rupture of the permeator. In our application, a fingerlike assembly of the membrane tubes inside the reactor has been designed: it permits the free elongation and contraction of the palladium alloy tube avoiding any mechanical cycling stress.

A process in which a membrane reactor produces ultra pure hydrogen needed to feed a polymeric fuel cell of power 500 W has been designed: the process foresees the use of both traditional and membrane reactors. The membrane reactor consists of a bundle of Pd-Ag thin wall tubes operating in parallel where the water gas shift reaction takes place at 350°C and 200 kPa: the hydrogen stream of $4{,}46\cdot10^{-3}$ mol s^{-1} (about 6 liter/min) permeating through the membrane is recovered in the shell side by means of nitrogen sweep gas at 100 kPa.

Keywords: Pd-Ag Membranes, Hydrogen Production, Membrane Reactor Design.

Introduction

Due to their property to be selectively permeable to hydrogen, palladium and its alloys are proposed for manufacturing membranes and membrane reactors in applications for separating and producing hydrogen [1-3]. Particularly, the ability of the membrane to separate hydrogen during dehydrogenation reactions involves the shift of the reaction conversion beyond the thermodynamical equilibrium values attainable in traditional reactors

Permeator tubes have been produced via diffusion welding of Pd-Ag foils of thickness 0.050 mm [4]. This procedure ensures higher hydrogen permeance and reduced costs with respect to commercial tubes of wall thickness 0.100 mm. However, as a consequence of the hydrogen loading into the palladium alloy, the permeators modify significantly their length and only an appropriate design of the membrane module could avoid mechanical stresses and ensure a long life of the membrane tubes.

Actually, in the framework of a project aimed at demonstrating the efficiency of an integrated system consisting of a reformer producing ultra pure hydrogen feeding a polymeric fuel cell of power 500 W, the production of hydrogen via ethanol reforming in a membrane process has been studied. In fact, the Pd-Ag thin wall membranes separate hydrogen with an infinite selectivity and therefore avoid the presence of impurities such as CO that may poison the polymeric fuel cells.

New Design of Pd-Ag tubular membrane reactors

The permeator tubes require an appropriate design of the membrane module in order to avoid also modest mechanical stresses that may damage the very thin wall membrane tube produced by diffusion welding. In fact, under thermal and hydrogenation cycling the palladium alloy the Pd-Ag tubes elongate and contract modifying significantly their length and form as reported in fig. 1 where a thin wall permeator is shown before and after testing with hydrogen.

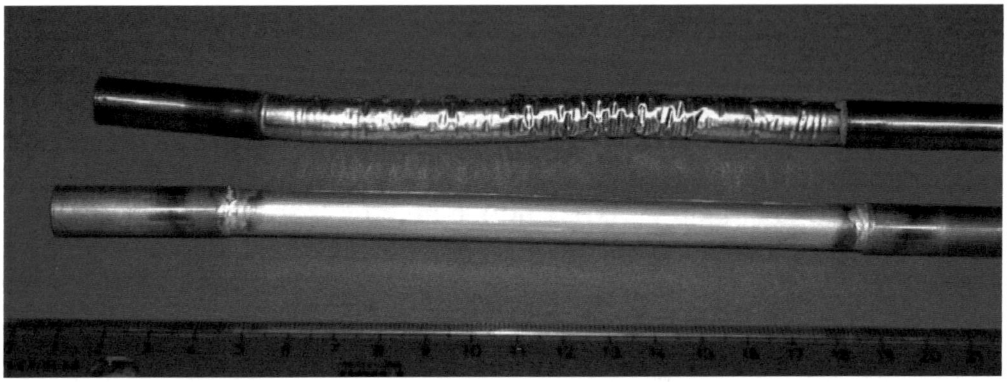

Figure 1: The Pd-Ag permeator tube after testing (above) compared to a virgin one (below).

Hence, improper mechanical constraints may disallow these strains producing a combined compression and bending stresses of the permeator and involving its failure. A new Pd-Ag membrane module using a fingerlike configuration has been designed for laboratory applications [5]: the permeator is free in its elongation or contraction and can change its length without rising stresses of the thin wall tube. This module can be used for testing dehydrogenation reactions. The case of a membrane reactor where the catalyst is located inside the Pd-Ag permeator tube is shown in fig. 2. The feed stream is sent inside the membrane through a steel tube: here, on the catalyst bed, it reacts producing hydrogen that is separated by the Pd-Ag and is collected by the sweep gas in the shell side.

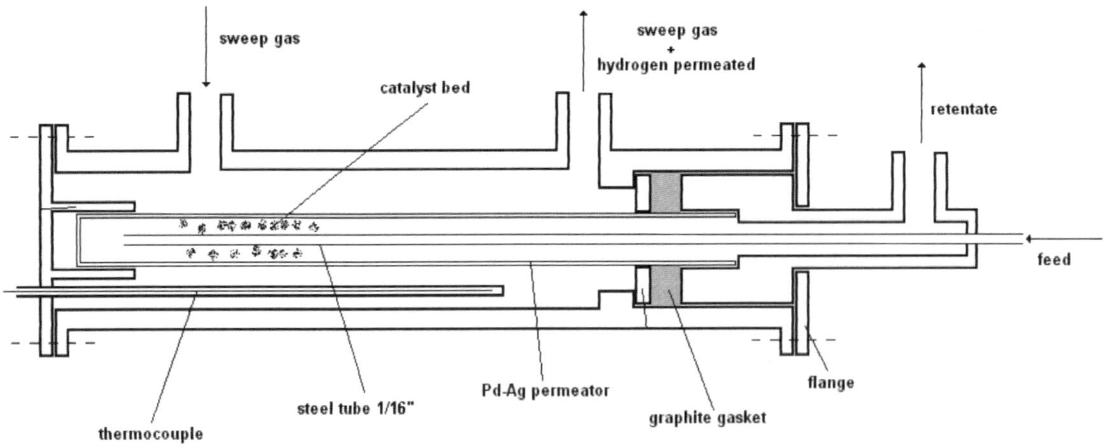

Fig. 2. Scheme of a membrane reactor with the catalyst bed inside the permeator.

Membrane ethanol reformer

Hydrogen is mainly produced by reforming of natural gas; other industrial applications use alcohols such as methanol and ethanol. Especially, the latter alcohol may be obtained by fermentation of biomass: the CO_2 produced during the reforming is the same absorbed by the plants from the environment via the chlorophyll synthesis. In this way, the process involving the use of hydrogen produced by biomass through ethanol reforming may be considered CO_2-neutral.
The ethanol reforming reaction is:

$$C_2H_5OH + H_2O \Leftrightarrow 2CO + 4H_2 \qquad \Delta H = 260 \text{ kJ mol}^{-1} \qquad (1)$$

The hydrogen production is maximized by the water gas shift reaction where the CO is converted into CO_2 and hydrogen:

$$CO + H_2O \Leftrightarrow CO_2 + H_2 \qquad \Delta H = -41.2 \text{ kJ mol}^{-1} \qquad (2)$$

Therefore, the complete reaction is:

$$C_2H_5OH + 3H_2O \Leftrightarrow 2CO_2 + 6H_2 \qquad \Delta H = 173{,}1 \text{ kJ mol}^{-1} \qquad (3)$$

The ethanol reforming is an endothermic reaction: typically, it is carried out at high temperature (600-800 °C). A wide bibliographic work on this reaction has been performed [6-8]: usually, catalysts based on Rh, Ru, Pd, Pt, Ni, Co and Cu are used on supports of Al_2O_3, SiO_2, MgO and La_2O_3. The reaction conversion and selectivity of the products (mainly H_2, CO, CO_2 and in minor part CH_4, CH_3CHO and C_2H_4) depend on the catalyst as well as on the temperature. A large excess of water in the feed stream increases the hydrogen selectivity and reduces the coke formation that may poison the catalyst particularly when the reaction is operated at low temperature.

The use of thin wall Pd-Ag permeators is not applicable in an ethanol membrane reformer as a consequence of the poor mechanical resistance of these devices at high temperature (600-800 °C). For this, a new process using two reactors has been studied [9]: a traditional reformer operating at high temperature where ethanol reacts with an excess of water is followed by a Pd-Ag membrane reactor in which the water gas shift reaction takes place, see fig. 3. In this second device the CO coming from the reformer reacts with the water yielding further hydrogen and completing the ethanol conversion to hydrogen: at the same time all the hydrogen produced in both reactors is selectively separated by the membrane and collected in the shell side of the membrane reactor.

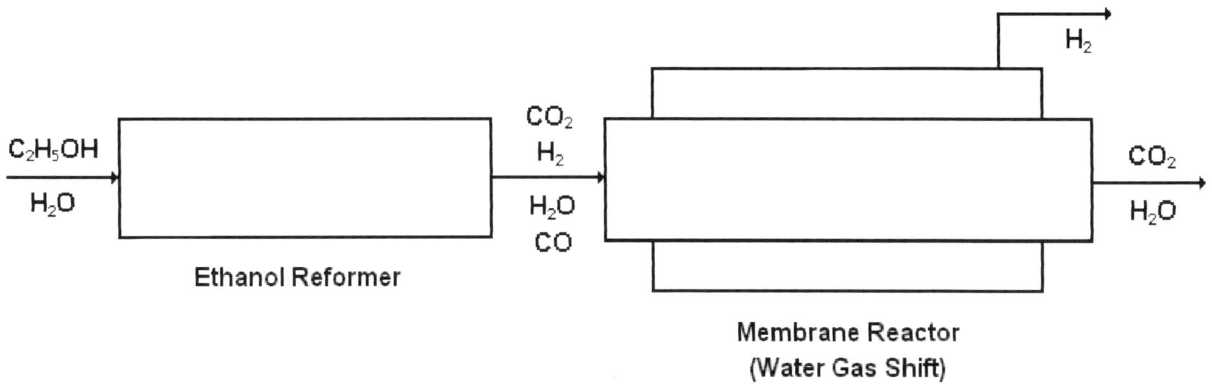

Fig. 3. Scheme of the two steps process

The process has been designed on the basis of the literature data reported for an ethanol reformer using a Rh catalyst supported on alumina and operating at 650 °C [10]. With a H_2O/C_2H_5OH feed molar ratio of 8.4 and a space velocity (GHSV) of 5000 h^{-1} are produced about 5.5 moles of hydrogen per each ethanol mole reacted: high selectivity of CO_2 (75%) and CO (23%) is attained with small amounts of minor products like methane (< 2%). In the proposed process the ethanol reformer is a traditional fixed bed tubular reactor operating at 650 °C and 200 kPa: the calculated composition of the gaseous stream leaving the reformer and feeding the membrane reactor are reported in Table 1.

Table 1: composition of the stream feeding the membrane reactor.

gaseous specie	molar fraction
CO_2	0,116
CO	0,036
H_2O	0,419
H_2	0,427
Others (mainly CH_4 and CH_3CHO)	0,002

In order to convert the CO into CO_2 and produce further hydrogen, the water gas shift reaction takes place in a second reactor operating at 350 °C and 200 kPa. This device operates both as a membrane reactor (i.e. producing and separating the hydrogen produced by the water gas shift reaction) and a permeator (i.e. separating the hydrogen produced in the first reactor by reforming the ethanol). This second reactor, see the scheme of fig. 4, uses a fingerlike configuration of 19 Pd-Ag tubes of diameter 10 mm, wall thickness 0.050 mm and length 300 mm: the hydrogen permeated is collected in the shell side by a nitrogen stream at 100 kPa in a contercurrent mode.

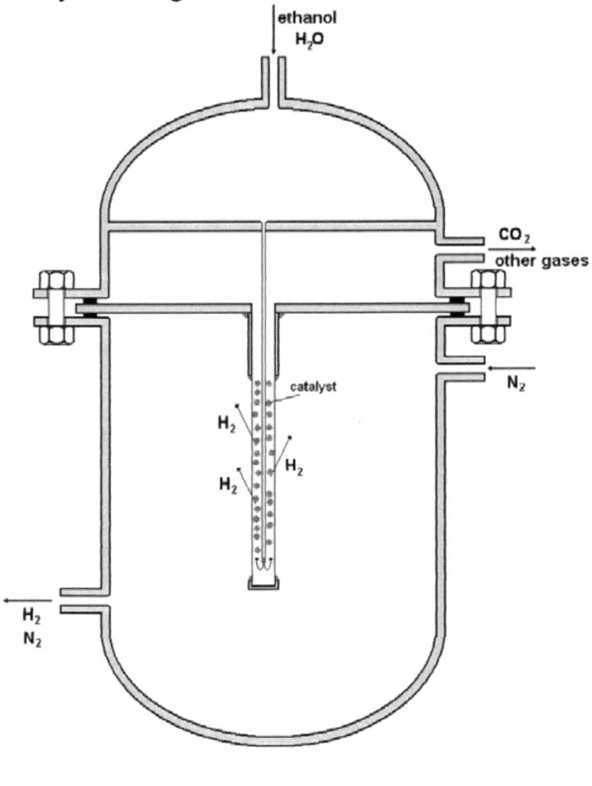

Fig. 4. Scheme of the multi tube Pd-Ag membrane reactor

The nitrogen purge stream operates in a loop including the anode of the polymeric fuel cell where the hydrogen reacts and the sweep nitrogen is sent back to the membrane reactor shell. The molar flowrates resulting from the mass balances are reported in Table 2 with reference to the process scheme of fig. 5.

Table 2: molar flow rate (mol s^{-1}) of the two steps process.

Stream	1	2	3	4	5	6	7	8	9	10	11	12	13	14
State	L	G	G	G	G	G	G	G	G	G	G	G	G	G
T, K	293,15	923,15	923,15	623,15	623,15	293,15	623,15	373,15	373,15	373,15	623,15	293,15	373,15	373,15
P, kPa	200,00	200,00	200,00	200,00	200,00	200,00	100,00	100,00	100,00	100,00	100,00	100,00	100,00	100,00
H_2O	21,00	21,00	15,06	15,06	13,70	13,70								14,61
N_2							20,00	20,00		20,00	20,00	31,17	31,17	31,17
O_2												8,04	8,04	0,73
H_2			13,25	13,25			14,61	14,61	14,61					
CO			1,37	1,37	0,01	0,01								
CO_2			3,50	3,50	4,86	4,86								
C_2H_5OH	2,50	2,50	0,06	0,06	0,06	0,06								

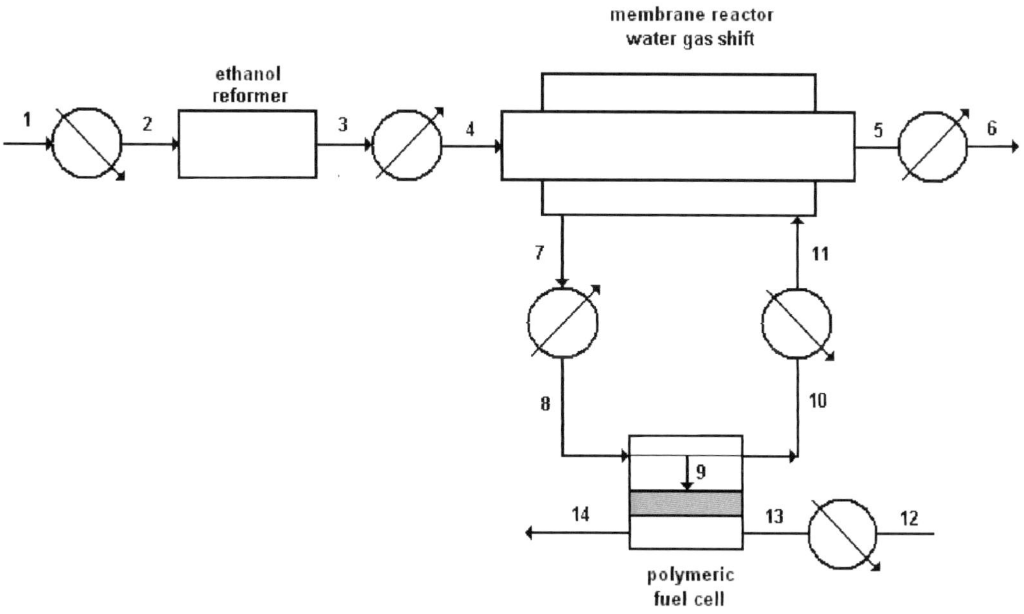

Fig. 5. Scheme of the system consisting of the membrane reformer and the polymeric fuel cell

The membrane reactor has been designed in order to maximize the hydrogen recovered in the shell side: in fig. 6 the calculated molar flow rates of the gaseous species inside a permeator tube of the membrane reactor are reported. All CO reacts with water producing hydrogen: this gas permeates almost completely the membrane into the reactor shell where is recovered by nitrogen.

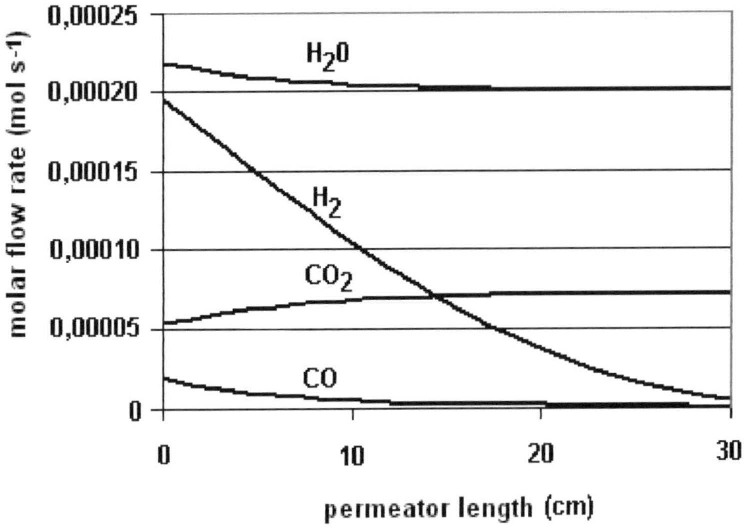

Fig. 6. Molar flow rate profiles inside the Pd-Ag tubes of the water gas shift membrane reactor

Discussion and Conclusions

Pd-Ag permeator tubes having a reduced wall thickness (0.050 mm) with respect to the commercial ones (0.100 mm) have been built via a diffusion welding procedure: they have been proposed for producing highly pure hydrogen in membrane reactors via ethanol reforming. These thin wall permeators have a high hydrogen permeance, infinite selectivity and reduced cost. Their reliability has been increased by adopting a fingerlike configuration inside the membrane module: such a kind of design avoids any mechanical stress of the permeator tube due to the Pd-Ag elongation under hydrogen loading and assuring a long lifetime.

A membrane process has been studied: the ethanol reforming is carried out in a traditional reactor while both the CO conversion and the hydrogen permeation take place in a multi-tube membrane reactor. Main characteristics of the proposed process are the complete conversion of ethanol, the production of pure hydrogen and the reduced cost. In fact, the membrane reactor consists of a bundle of 19 Pd-Ag tubes: each tube uses about 5 g of palladium for a total cost of about 700.00 € (estimated palladium cost of 8.00 €/g). This value could be considered acceptable if compared to the costs of the polymeric fuel cell of power 500 W which would be coupled to the membrane reactor.

Resuming, the high hydrogen permeability and the long duration of Pd-Ag thin wall permeator tubes besides of the reduced costs make this membrane technology available for producing highly pure hydrogen in energetic and industrial applications (PEM fuel cells and automotive systems, chemical industry, etc.).

References

[1] Shu, J., Grandjean, B.P.A., Van Neste, A., Kalaguine, S., Catalytic palladium-based membrane reactors: a review, Canadian Journal of Chemical Engineering, vol. 69, Oct. 1991, 1036-1060

[2] Kikuchi, E., Membrane reactor application to hydrogen production, Catal Today 56 (2000) 97–101

[3] Armor, JN., Applications of catalytic inorganic membrane reactors to refinery products, Journal of Membrane Science 147 (1998) 217–33

[4] Tosti, S., Bettinali, L., Lecci, D., Marini, F., Violante, V., Method of bonding thin foils made of metal alloys selectively permeable to hydrogen, particularly providing membrane devices, and apparatus for carrying out the same, European Patent EP 1184125
[5] Basile, A., Tosti, S., Dispositivo portatile a membrana intercambiabile per valutazione di processi di permeazione e reazione in fase gassosa, Italian Patent n. RM2005U000107, 2005
[6] Liguras, D.K, Kondarides, D.I., Verykios, X.E, Production of hydrogen for fuel cells by steam reforming of ethanol over supported noble metal catalysts, Applied Catalysis B: Environmental, 43 (2003) 345-354.
[7] Breen, J.P., Burch, R., Coleman, H.M., Metal-catalysed steam-reforming of ethanol in the production of hydrogen for fuel cell applications, Applied Catalysis B: Environmental, 39 (2002) 65-74.
[8] Llorca, J., Homs, N. , Sales, J. , De La Piscina, P.R., Efficient Production of Hydrogen over Supported Cobalt Catalysts from Ethanol Steam reforming, Journal of Catalysis,209 (2002) 306-317
[9] Tosti, S., Basile, A., Bettinali, L., Chiaravalloti, F., Study and characterization of Pd-based permeators used in membrane reformers, Proc. International Hydrogen Energy Congress and Exhibition IHEC 2005, Istanbul, Turkey, 13-15 July 2005
[10] Cavallaro, S., Chiodo, V., Freni, S., Mondello, N., Frusteri, F., Performance of Rh/Al_2O_3 catalyst in the steam reforming of ethanol: H_2 production for MCFC, Applied Catalysis A: General 249 (2003) 119–128

99f High-Pressure Co2/Ch4 Separation Using Sapo-34 Membranes

Richard D. Noble[1], John L. Falconer[1], Shiguang Li[1], Janna Martinek[1], and Tracy Q. Gardner[2]. (1) Chemical and Biological Engineering, University of Colorado, 1111 Engineering Drive, Boulder, CO 80309-0424, (2) Chemical Engineering, Colorado School of Mines, Golden, CO 80401

SAPO-34 membranes were prepared by in-situ crystallization onto tubular stainless steel supports (0.8-μm pores). The SEM image of the surface of membrane M2 shows intergrown zeolite crystals that are rectangular with the approximate dimension of (50-100) × (100-200) nm. The SEM cross section of the membrane indicated that the continuous zeolite layer was approximately 5-μm thick. The n-C4H10 permeance at 473 K for membrane M1 was below the detection limit of 1 × 10-12 mol/(m2•s•Pa). Both CO2 and CH4 single-gas fluxes through membrane M1 increased with pressure drop at all h), which is more temperatures used. The highest CO2 flux measured was 45 kg/(m2 than an order of magnitude higher than most pervaporation fluxes through zeolite membranes. Carbon dioxide permeates faster than CH4 because the smaller CO2 diffuses faster and because it has higher adsorption coverages than CH4.The CO2 fluxes decreased as the temperature increased except at high pressures for the temperature increase from 297 to 373 K. The CH4 fluxes were almost independent of temperature. The membranes were used for separating CO2 from CH4 at feed s•Pa) pressures up to 3.1 MPa. The highest CO2 permeance was 2.4 × 10-7 mol/(m2 for a 50/50 feed mixture at a pressure drop of 0.14 MPa. For a pressure drop of 3 MPa, the CO2/CH4 separation selectivities at 253 K were 140 – 150; at lower pressure drops, the highest selectivity was 270. The highest CO2 flux was 21 h) at 295 K and a pressure drop of 3 MPa. Both CO2 and CH4 fluxes in an kg/(m2 equimolar CO2/CH4 mixture increased with feed pressure at 295 K when the permeate pressure was kept constant at 84 kPa. The separation selectivity increased slightly as feed pressure increased because the permeance decreased proportionally more for CH4 than CO2. The separation selectivity was as much as a factor of 3 higher than the ideal selectivity, with the largest difference seen at high pressure drops. The CH4 coverage is lower in the mixture, relative to the single gas, because of competitive adsorption; CO2 has a higher heat of adsorption (24 kJ/mol) than CH4 (16 kJ/mol) in the SAPO pores. The

M-S model was used to predict how CO_2 and CH_4 fluxes changed with pressure. The adsorption equilibrium constants measured for SAPO crystals were used. The CO_2 and CH_4 single-gas fluxes at low pressure were used to estimate the CO_2 and CH_4 diffusivities. The model indicates that the CO_2 flux at the same CO_2 pressure is lower in the mixture than for the single gas because the slower-diffusing CH_4 slows the faster diffusing CO_2. The model also indicates that the CH_4 flux is lower in the mixture than for the single gas because CO_2 inhibits CH_4 adsorption, and this more than compensates for the increased CH_4 diffusivity in the presence of CO_2. The CO_2 mixture fluxes from the model are in reasonable agreement with the measurements. The CH_4 mixture fluxes from the model, however, are not in agreement with the measurements, because a significant fraction of the CH_4 flux is through non-zeolite pores. When the feed pressure was kept constant at 3.1 MPa for an equimolar CO_2/CH_4 mixture and the permeate pressure was increased, both CO_2 and CH_4 fluxes decreased as the driving force for permeation decreased. Because the permeate was enriched in CO_2, the partial pressure difference (and thus the driving force) across the membrane decreased more for CO_2 than CH_4 as the permeate pressure increased. Thus, the CO_2 flux decreased proportionally more than the CH_4 flux. Therefore, the CO_2 permeate concentration decreased as the permeate pressure increased. The measured CO_2 flux dependence on permeate pressure was in close agreement with that predicted from the M-S model. When the CO_2 feed concentration was increased at 253 K, for a pressure drop of 3.0 MPa across membrane M1, the CO_2 flux increased and the CH_4 flux decreased. Thus, the CO_2 permeate concentration increased. Even for a 25% CO_2 feed, the CO_2 permeate concentration was 94.6%. As the CO_2 feed concentration increased from 25 to 70%, the CO_2 adsorption coverage on the feed side increased from 65% to 92% of saturation, based on the mixture isotherm. The CO_2/CH_4 selectivity exhibited a maximum of 180 at a CO_2 feed concentration of 50%.

99g Novel Thermal Swing Sorption Enhanced Reaction (Tsser) Process Concept for Hydrogen Production

Hugo S. Caram, Alexander Verdooren, and Shivaji Sircar. Chemical Engineering, Lehigh University, 111 Research Drive, Bethlehem, PA 18015

A novel thermal swing sorption enhanced reaction (TSSER) concept is being developed for direct production of essentially carbon oxide free hydrogen by low temperature (~400- 450 C) steam-methane reforming (SMR). The concept removes the reaction by- product, CO_2, from the reaction zone of the reactor by a CO_2 selective chemisorbent (K_2CO_3 promoted hydrotalcite). This paper describes the CO_2 sorption equilibria on the chemisorbent at different temperatures, characterisitics of the CO_2 mass transfer zones on the chemisorbent column under various operating conditions, and the isothermal as well as thermal desorption characteristics of CO_2 from the chemisorbent. Column dynamic and gravimetric sorption methods are used to generate the data. Appropriate thermodynamic and kinetic models are used to describe the data. The possible mechanism of the surface reactions leading to the chemisorption of CO_2 based on surface analysis by Raman spectroscopy is discussed.

142a Adsorption and Membrane Processes in Hydrogen Production

James A. Ritter and Armin D. Ebner. Department of Chemical Engineering, University of South Carolina, Columbia, SC 29208
A Chemical Industry Vision 2020 sub-committee was established in June 2004 and chartered with the responsibility of identifying research and development needs for separation technology to drive equilibrium processes for the Chemical and Petrochemical Industry. This study focused on the anticipated drivers and requirements of the chemical and petrochemical industries. Chemical production processes were selected for review. This report, the first in a series, is directed to commercial, industrial hydrogen production. The growing requirements for hydrogen in chemical manufacturing, petroleum refining and the new emerging clean energy concepts will place greater demands on sourcing, production capacity and supplies.

A review was conducted of the current state of the art and emerging literature concepts on both adsorption and membrane separation technology applicable to H2 production. Recommendations for future R&D needs are discussed. An emphasis is placed on flow sheet design modification with adsorption or membrane units being added to existing plants for near term impact, and on new designs with complete flow sheet modification for new adsorption or membrane reactor/separators replacing current reactor and separator units in an existing plant for a longer term sustainable impact.

About 41 MM tons/yr of H2 is produced worldwide, with 80% of it being produced from natural gas by steam reforming, partial oxidation or autothermal reforming. H2 is used commercially to produce CO, syngas, ammonia, methanol and higher alcohols, urea and hydrochloric acid. It is also used in Fischer Tropsch reactions, as a reducing agent (metallurgy), and to upgrade petroleum products and oils (hydrogenation). Current industrial practices are summarized in terms of the key reforming and shift reactions and reactor conditions, with the four most widely used separations techniques --absorption, adsorption, membrane and cryogenic. As all of these reactions are reversible, the H2 or CO or syngas production from natural gas and hydrocarbon feedstocks is equilibrium limited and consequently provides opportunities for process improvement

It has been estimated that the reforming of natural gas to produce H_2 consumes about 31,800 Btu/lb of H_2 produced. It is further estimated that 450 trillion Btu/yr could be saved with a 20% improvement in just the H_2 separation and purification train after the H_2 reformer. This energy savings is a major driver for future R&D and for implementation of adsorption or membrane separation technology.

Recommendations for future separations R&D are set forth in – Adsorbent Materials Development, Membrane development, Plant Design modification with additional separations systems, and Plant systems optomization. Detailed performance requirements are provided in the report. An overview of this report will be given during this presentation.

142b Chemically Crosslinked Polymer Nanocomposite Membranes for Hydrogen Separation

Christopher Aberg[1], Jon Samseth[2], May-Britt Hagg[3], and **Richard Spontak**[1]. (1) Chemical & Biomolecular Engineering, North Carolina State University, 911 Partners Way, Raleigh, NC 27695, (2) Chemistry & Materials, SINTEF, Trondheim, Norway, (3) Chemical Engineering, NTNU, Trondheim, Norway

Gas separation technologies have witnessed a number of advances over the past decade as functionalized materials are being designed on the basis of evolving material design paradigms. Polymeric membranes, in particular, are undergoing a renaissance as a low-cost, energy-efficient, environmentally-friendly means of separating gases on the basis of both size and solubility. As nanostructured polymers continue to develop, many of the insights gained from these multifunctional materials can be exploited in the molecular-level design of selective membranes. Recent efforts have demonstrated that the incorporation of dispersed nanoparticles within a polymer matrix can alter the local free volume of polymeric membranes, as well as improve the affinity for one penetrant gas species over another. Moreover, chemical cross-linking of such membranes can further influence the available free volume of a polymer, selectively altering the diffusion of larger gas molecules. Some of our previous findings related to H2 purification from mixed CO2/H2 gas streams will be briefly reviewed. In the present work, we present a straightforward strategy for developing cross-linked H2-selective polymers containing palladium (Pd) nanoparticles grown in-situ. Previous work has shown that such cross-linking can greatly reduce the permeabilities of larger gases (e.g., CO2, O2 and N2) without hindering the permeabilities of smaller gases (e.g., He and H2), thereby increasing the ideal selectivity of H2 relative to larger gases, in some instances, above the Robeson trade-off line. Nanoparticles of Pd are integrated into the polymer membrane to not only reduce the free volume available for molecular transport, but also increase the solubility of H2. We will present chemical evidence for cross-linking of the polymers under investigation and microscopic evidence for the incorporation and distribution of Pd nanoparticles. Pure- and mixed-gas permeability and sorption data will be reported to illustrate the effects of chemical cross-linking and Pd nanoparticles on H2 permeability and H2/gas selectivity relative to other polymeric membranes.

142c Sorption-Enhanced Reaction Processes: Steam Methane Reforming Combined with in-Situ Co2 Removal for Increased Hydrogen Production

Alírio E. Rodrigues, Department of Chemical Engineering, University of Porto, LSRE - Laboratory of Separation ad Reaction Engineering, LSRE - FEUP Rua Dr. Roberto Frias, Porto, Portugal

Sorption-Enhanced Reaction Processes: Steam Methane Reforming Combined With in-situ CO_2 Removal for Increased Hydrogen Production.

Eduardo L. G. Oliveira, Carlos A. Grande and Alírio E. Rodrigues *

LSRE – Laboratory of Separation and Reaction Engineering, Department of Chemical Engineering, Faculty of Engineering, University of Porto. Rua Dr. Roberto Frias (4200-465), Porto, Portugal.

New global policies on energy production are leading to stringent process optimisation and development. Attending to reduce the emission of greenhouse gases, new processes and new fuels are under research. Hydrogen is envisioned as the fuel of the future. Actually, almost 80% of hydrogen consumed worldwide is produced by Steam Methane Reforming (MSR).

The most important reactions in the MSR are:

$$CH_4 + H_2O \Leftrightarrow CO + 3H_2 \qquad \Delta H_{298K} = 206 kJ/mol \qquad (1)$$

$$CO + H_2O \Leftrightarrow CO_2 + H_2 \qquad \Delta H_{298K} = -41 kJ/mol \qquad (2)$$

$$CH_4 + 2H_2O \Leftrightarrow CO_2 + 4H_2 \qquad \Delta H_{298K} = 165 kJ/mol \qquad (3)$$

All these reactions are reversible and the overall scheme results endothermic. As can be noted from the reaction scheme, this reaction is thermodynamically limited and requires very high temperatures to obtain reasonable methane conversions, increasing energy demands of the process and also capital cost. The removal of products from the reaction media (removal of hydrogen and/or carbon dioxide) can displace the equilibrium to obtain higher amounts of hydrogen.

The removal of hydrogen is being studied through the use of membrane-aided reformers, where H_2 is removed from the media shifting equilibrium and at the same time obtaining a purified stream. The second route for enhance hydrogen production by equilibrium shifting is the removal of CO_2 from the reaction media.

In our group we have been studying CO_2 adsorption at high temperatures using hydrotalcites. An example of the amount adsorbed of CO_2 at 673K in hydrotalcite (Condea) measured by breakthrough curves is shown in Figure 1. The removal of carbon dioxide from the reaction media has two advantages: it displaces reaction equilibrium resulting in higher yields of hydrogen and also produces a stream with less CO (undesirable by-product for fuel cells). This work reports experimental data of a large-pore Ni-based catalyst (see SEM picture in Figure 2) combined with high temperature adsorption data of CO_2 in hydrotalcite. The combination of MSR reaction – CO_2 adsorption (SERP concept) is also presented.

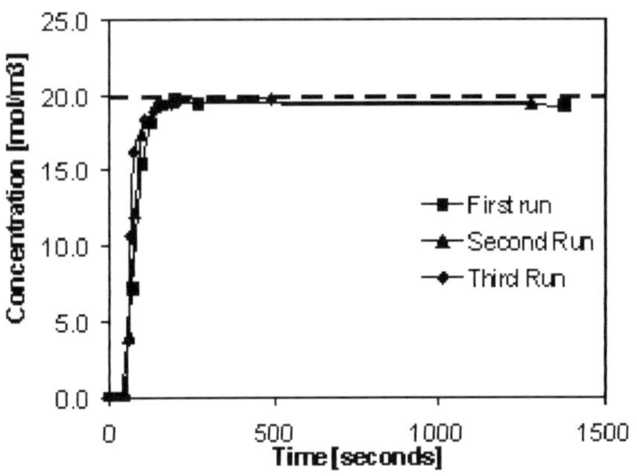

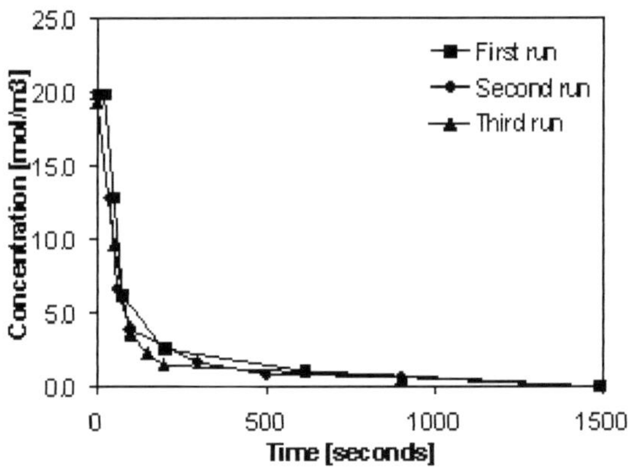

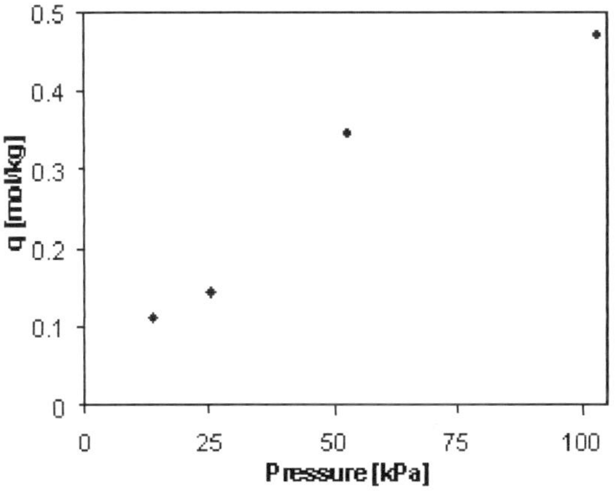

Figure 1. Adsorption of CO_2 on hydrotalcite at 673K: breakthrough curves (adsorption / desorption cycles) and pure component equilibrium.

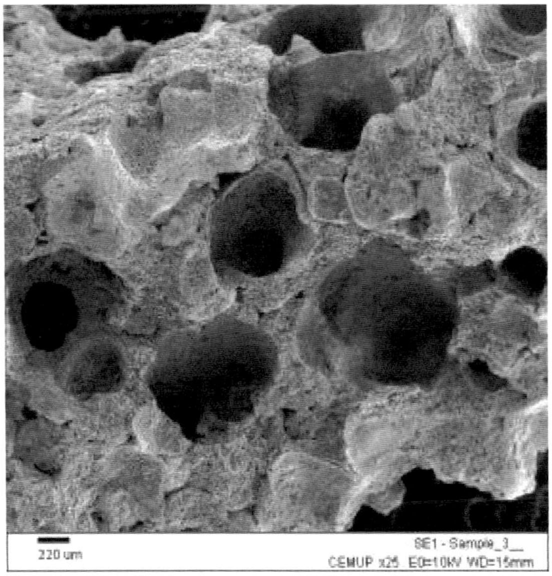

Figure 2. SEM picture of large pore Ni-based catalyst used for steam methane reforming.

Methanol steam reforming and Ethanol steam reforming in Membrane Reactors: an experimental study

A.Basile[1], A. Iulianelli[2], F. Gallucci[1], S. Tosti[3], E. Drioli[1,2]*

[1] *Institute on Membrane Technology, ITM-CNR, c/o University of Calabria, via P. Bucci, Cubo 17/C, 87030, Rende (CS), Italy, Phone: (+39) 0984 492013 – Fax: (+39) 0984 402103 –a.basile@itm.cnr.it*

[2] *Department of Chemical Engineering and Materials, University of Calabria, via P. Bucci Cubo 42/A, 87030 Rende(CS), Italy, Phone: (+39) 0984 492039 – Fax: (+39) 0984 402103*

[3] *ENEA, Unità Tecnico Scientifica Fusione, C.R. ENEA Frascati, Via E. Fermi 45, Frascati, Roma 00044, Italy, Phone: +39 06 94005160 – Fax: +39 06 94005374*

ABSTRACT

In this work a comparison between methanol steam reforming reaction (MSR) and ethanol steam reforming reaction (ESR) to produce hydrogen in membrane reactors (MRs) is discussed from an experimental point of view .

Both reaction systems have been investigated by considering the influence of the membrane characteristics as well as the influence of the operating temperature. In the case of the dense membrane, the sweep gas flow rate and the different fluxes configurations have also been analysed.

Experimental results, in terms of reactants conversion as well as hydrogen production and gases selectivity in MRs and in a traditional reactor (TR), are presented. The catalyst stability in both reaction systems was also tested.

Keywords: Methanol, Ethanol, Steam Reforming, Hydrogen, Catalyst, Membrane Reactor,

1. INTRODUCTION

The polymeric fuel cell systems are electrochemical devices able to generate electrical power by electrochemical oxidation of hydrogen with atmospheric oxygen. Many studies about hydrogen production for fuel cells are dealing with the use of two types of carbon compounds: the first one is an oxygen-containing compounds, such as methanol or ethanol, while the second one is represented by hydrocarbons, such as natural gas, propane gas, gasoline, etc. [1-8].

Among the different fuels, the advantages of both methanol and ethanol include their easiness to handle and to transport than gases and their low cost: furthermore, these compounds can be synthesized from a lot of feedstock's (biomass, coal and natural gas, etc) under mild conditions in the reactions for producing hydrogen. The disadvantages include the relatively high concentration of carbon monoxide in the reaction products, which is still much higher than the limit of tolerance of carbon monoxide concentration in feed gas for polymeric electrolyte membrane (PEM) fuel cells.

The aim of this experimental work has been the investigation of the steam reforming of ethanol and methanol in palladium-based membrane reactors by using a commercial 5% Ru/Al_2O_3 catalyst that can be used for both the reactions. The influence of different parameters such as the membrane type, the temperature, the sweep gas configuration and the sweep gas flow rate was investigated.

The results, in terms of reactants conversion and hydrogen production, are presented as a comparison between the steam reforming reactions of methanol and ethanol.

2. Description of the process

2.1. Traditional and membrane reactors description

The TR consists of a stainless steel tube, length 250 mm, *i.d.* 6.7 mm with the reaction zone 150 mm. The MR consists of a tubular stainless steel module, length 280 mm, *i.d.* 20 mm, containing different membranes, the first one (MR1) is a TiO_2-Al_2O_3 asymmetric porous commercial membrane (Inoceramic) with a Pd-Ag deposit. The second membrane (MR2) is a commercial asymmetric porous ceramic membrane with a Pd-Ag deposit. The third membrane (MR3) is a pine-hole free Pd-Ag thin wall membrane tube permeable only to hydrogen, having thickness 50 μm, *o.d.* 10 mm, length 150 mm. In particular, the dense membrane is joined to two stainless steel tube ends useful for the membrane housing. In the MRs, catalyst pellets are packed in the membrane zone (150 mm length) while glass spheres (2 mm diameter) are placed into the supports on both extremities of the membrane.

2.2. Experimental details

The reactor (TR or MRs) is placed in a temperature-controlled P.I.D. (Proportional + Integral + Derivative Control) oven. Reaction and permeation temperatures are in the range between 350 and 600°C for MR1 and MR2, while the temperature range is 350-450°C for MR3. The sweep gas (N_2) is fed (only for MR3) by means of a mass-flow controller. H_2O and methanol or ethanol are fed by means of volumetric pumps.

The reaction pressure is held up to 1.3 bar by means of a regulating-valve system placed at the outlet side. Considering the MR, permeate pressure is always 1 bar, and N_2 is used as the sweep gas (MR3) with the flow rate in the range between 0.5 and 12 mL/s. The methanol or ethanol feed flow rates are $1.278 \cdot 10^{-4}$ mol/s while the H_2O/other reactant feed ratio is 4.5. The outlet streams are completely condensed in order to remove the un-reacted H_2O and the alcohol, and then the liquid phase is analysed by means of a Abbe Refractometer at 25°C. The dry gaseous stream flow rate is measured by means of bubble flow-meters; its composition is detected by using a temperature programmed HP 6890 Gas Chromatograph (GC) with a TCD (Thermal Conductivity Detector) at 250°C and He as carrier gas. All the TR and the MRs were packed with 3 g of a 5%wt Ru-Al_2O_3 commercial catalyst furnished by Johnson Matthey. Before reaction, the catalyst was pre-heated using N_2 at 400°C under atmospheric pressure for 3 hours and, afterwards, reduced by using H_2 ($1.5 \cdot 10^{-3}$ mol/min) at the same temperature for 2 hours.

3. Results and discussion

The experimental results are presented as a comparison between the methanol steam reforming and the ethanol steam reforming for each membrane reactor. The best results will be emphasized in the conclusions of the work.

All the experimental results reported in the present work are average values of different experimental points (at least three) taken in a fixed time (2 hrs). The balance on the C is closed with a 4% error.

In order to evaluate the performance of the reactor, an important parameter to be considered is the total hydrogen production. Table 1 summarizes the hydrogen production by MSR and ESR in both TR and MR1. Considering the MSR reaction, for temperatures higher than 450°C, the hydrogen production in the MR1 is higher than the hydrogen production in the TR. Moreover, the MSR gives a hydrogen production always higher than the ESR in both TR and MR1. Considering the ESR reaction, in the same table it can be seen that the hydrogen production in MR1 is higher than the hydrogen production in TR for the whole range of temperatures considered.

Table 1. Hydrogen production (mol/s) in MR1 and TR for ESR and MSR reaction.

Temperature °C	TR (ESR)	MR1 (ESR)	TR (MSR)	MR1 (MSR)
350	2,63E-05	8,46E-05	1,68E-04	8,20E-05
400	3,88E-05	1,04E-04	1,95E-04	1,61E-04
450	6,28E-05	1,18E-04	1,99E-04	1,99E-04
500	7,31E-05	1,44E-04	2,12E-04	2,42E-04
550	8,50E-05	1,57E-04	2,27E-04	2,81E-04
600	9,40E-05	1,69E-04	2,47E-04	2,81E-04

Also for MR2 the total hydrogen production (Table 2) follows the behaviour of the reactants conversion, that is to say the hydrogen production is increasing with the temperature for both ESR and MSR reactions in the MR2 and TR. Concerning the MSR reaction, the hydrogen production in MR2 is much lower than the hydrogen production in TR, for instance $1.43{\cdot}10^{-4}$ mol/s in MR2 and $2.27{\cdot}10^{-4}$ mol/s in TR at T=550°C. As far as ESR reaction is concerning, the results in terms of hydrogen production are more interesting. In fact, although the ethanol conversion in MR2 is always lower than the ethanol conversion in TR, the hydrogen production for temperatures lower than 450°C in MR2 is higher than the hydrogen production in TR. This is the only improvement that MR2 can give for the reaction systems considered.

Table 2. Hydrogen production (mol/s) in MR2 and TR for ESR and MSR reaction

Temperature °C	TR (ESR)	MR2 (ESR)	TR (MSR)	MR2 (MSR)
350	2,63E-05	4,66E-05	1,68E-04	7,90E-05
400	3,88E-05	5,34E-05	1,95E-04	8,82E-05
450	6,28E-05	5,96E-05	1,99E-04	1,04E-04
500	7,31E-05	6,81E-05	2,12E-04	1,25E-04
550	8,50E-05	7,59E-05	2,27E-04	1,43E-04

The membrane reactor MR3 uses a dense Pd-Ag membrane with a perm-selectivity H_2/(other gases) infinite. For this membrane reactor there are at least two advantages: the first one is that the shell stream is CO-free and so it can be directly fed to a PEMFC. The second advantage is that, in order to improve the hydrogen recovery in this reactor, a sweep gas can be used in the shell side.

Different flow rates of the sweep gas can be used both in co-current and in counter-current mode.

Considering the hydrogen production in MR3 it can be seen in Table 3 that, for the ESR reaction, the MR gives a higher hydrogen production than TR both when MR is operated in co-current mode and in counter-current mode. Concerning the MSR reaction, the MR in counter-current mode gives a higher hydrogen production than the TR. In each case, the MSR reaction gives higher hydrogen production than the ESR reaction.

Table 3. Hydrogen production (mol/s) in MR3 and TR for ESR and MSR reaction.

Temperature °C	TR (ESR)	MR3 (ESR) Co-Current	MR3 ESR) Counter-Current	TR (MSR)	MR3 (MSR) Co-Current	MR3 (MSR) Counter-Current
350	2,63E-05	3,32E-05	4,38E-05	1,68E-04	1,64E-04	1,76E-04
400	3,88E-05	8,04E-05	8,48E-05	1,95E-04	1,94E-04	2,37E-04
450	6,28E-05	1,65E-04	1,80E-04	1,99E-04	2,15E-04	3,68E-04

4 Conclusions

The methanol and ethanol steam reforming reactions have been studied from an experimental point of view in TR and in three different MRs packed with commercial Ru-based catalyst. Two Pd/Ag supported membranes (MR1 and MR2) and a dense Pd-Ag membrane (MR3) were used in order to increase the reactant conversion and the hydrogen production. The experimental results show that in both traditional and membrane reactors

MSR conversions are higher than ESR. For MSR, the highest conversion is achieved in MR3 where 100% conversion is reached at T = 400°C, in counter-current mode and 11.8 mL/s sweep gas flow rate.

For ESR, the highest conversion is achieved in MR3 at T = 450°C, in counter-current mode and sweep gas flow rate 11.8 mL/s, and it is equal to 50%. Also the total molar hydrogen flow rates, in both traditional and membrane reactors, for MSR are higher than ESR and the best results are obtained in MR3. The experiments with each membrane were performed in around 3 months. After these experiments, the three membranes showed the initial permeability and selectivity values. In particular, the membrane MR3 showed a permselectivity H_2/(other gases) infinite.

REFERENCES

[1] P.J. de Wild and M. J. F. M. Verhaak, Catalytic production of hydrogen from methanol, Catal. Today, 60 (2000) 3-10

[2] W. Wiese, B. Emonts and R. Peters, Methanol steam reforming in a fuel cell drive system, J. Power Source, 84 (1999) 187-193

[3] J. P. Breen, R. Burch and H. M. Coleman, Metal-catalysed steam reforming of ethanol in the production of hydrogen for fuel cell applications, Appl. Catal. B: Environmental, 39 (2002) 65-74

[4] S. Cavallaro, V. Chiodo, A. Vita and S. Freni, Hydrogen production by auto thermal reforming of ethanol on Rh-Al_2O_3 catalyst, J. Power Source, 123 (2003) 10-16

[5] G. Maggio, S. Freni and S. Cavallaro, Light alcohols/methane fuelled molten carbonate fuel cells: a comparative study, J. Power Source, 74 (1998) 17-23

[6] A. Heinzel, B. Vogel, P. Hubner, Reforming of natural gas-hydrogen generation for small scale stationary fuel cell systems, J. Power Source, 105 (2002) 202-207

[7] J.P. Shen and C. Song, Influence of preparation method performance of Cu/Zn-based catalysts for low-temperature steam reforming and oxidative steam reforming of methanol for H_2 production for fuel cells, Catal. Today, 77 (2002) 89-98

[8] Bard Lindstrom and Lars J. Pettersson, Hydrogen generation by steam reforming of methanol over copper-based catalysts for fuel cell applications, Int. J. Hydrogen En., 26 (2001) 923-933

142e Use of Room Temperature Ionic Liquids for Carbon Dioxide Separations

Richard D. Noble[1], Carl A. Koval[2], Douglas Gin[3], Dean Camper[3], and Jason Bara[3]. (1) Chemical and Biological Engineering, University of Colorado, 1111 Engineering Drive, Boulder, CO 80309-0424, (2) Chemistry and Biochemistry Dept., University of Colorado, UCB 215, Boulder, CO 80309-0215, (3) University of Colorado, 1111 Engineering Drive, Boulder, CO 80309-0424

Room Temperature Ionic Liquids (RTILs) are solvents consisting entirely of ions resembling the ionic melts of metallic salts; however, RTILs are liquids at much lower temperatures (298 K). RTILs have unique properties including high thermal stability, high ionic conductivity, negligible vapor pressure, and non-flammability. These properties make RTILs potential benign solvent replacements for volatile organics traditionally used in organic synthesis and separation processes. However, future process development using RTILs will require an ability to predict gas solubilities and diffusion coefficients in RTILs since many reactions and separations involve permanent or condensable gases. We have developed an experimental apparatus that can measure both the solubility and diffusion coefficient of various permanent gases and organic vapors in RTILs. Regular Solution Theory has been successfully used to analyze and predict the solubility behavior using immidizolium-based RTILs. The diffusion coefficient is determined by the use of a semi-infinite permeation analysis. The Stokes-Einstein relationship is adequate to predict the diffusion coefficients. Results indicate that the separation selectivity is dominated by solubility selectivity. These results can be used to determine the best RTIL solvent for a given gas as well as an "upper bound" line on a Robeson plot for ionic liquids to compare to polymer membranes for a given separation. We have successfully incorporated RTILs into polymer matrices and produced mechanically stable films. This approach holds the promise to obtain "tunable" properties for composite structures for a given separation. One approach is to use hydrophilic polymers to prepare a homogeneous composite membrane. We have incorporated up to 70% ionic liquid into these structures and produced stable films. Another approach is to use ordered liquid crystal structures and incorporate the ionic liquid into the hydrophilic regions. These composite structures are being evaluated for a number of separations, including CO2 separation. CO2 separations are especially attractive since CO2 is very soluble compared to other gases and the separation is based on solubility and not diffusion differences. Commercial polymer membranes use diffusion differences as the primary separation mechanism.

Gas Separation in Coal Gasification

[1]J. C. Diniz da Costa, [2]D. Sholl, [3]G. Reed and [3]K. Thambimuthu

[1]ARC Centre for Functional Nanomaterials, School of Engineering, The University of Queensland, Brisbane Qld 4072, Australia
[2]Department of Chemical Engineering, Carnegie Mellon University, Pittsburgh, PA 15213 USA
[3]Centre for Low Emission Technology, Pullenvale Qld 4069, Australia

Abstract This work addresses potential advancements in integrated gasification combined cycle (IGCC) system, in particular focusing on membranes for gas separation at high temperature. These requirements can be met by metal and ceramic membranes and are reviewed hereinafter. High temperature membrane technology for gas separation is embryonic at this stage. The technology is attracting a large interest by research groups and is currently supported by major funding agencies around the world. Several key areas of interest include novel materials, fabrication process and membrane systems. Novel materials are required to address trace contaminants in dirty syngas processing, in particular the development of palladium alloys to tolerate sulphur. Also water has detrimental effects in many ceramic membranes, and functionalisation of silica surfaces delivers hydrostability. Fabrication processes to reduce film thickness and defects are taking ceramic membrane fluxes to 135 litres (STP) of H_2 per square metre of membrane area per minute, and at 1 atmosphere differential pressure across the membrane, and H_2/CO_2 selectivity in excess of 50. Palladium membranes show lower fluxes, but H_2/CO_2 selectivities are several orders of magnitude higher than that for ceramic membranes. Data on the performance of scale up and long term stability of membrane systems integration in IGCC is still scarce, whilst current costs of over US$ 1,000/m^2 of membrane area need to be reduced. From a technological view point, there is no single technology at the moment that can address all the future needs of syngas processing. It is likely that several technologies will address specific functions and requirements for gas separation in coal gasification.

Introduction

Security of fuel supply is becoming a major world issue, with key world oil reserves located in socially and politically sensitive areas of the globe. Further, great pressures are being exerted on energy intensive economies due to climate changes attributed to greenhouse gases, in particular CO_2 emissions. As demand for energy is unlikely to reduce, there has been a major technological shift worldwide towards more efficient and clean energy delivery systems. While oil and gas reserves are estimated to last for another 75 and 100 years, respectively [1], coal is a primary energy source, and in many countries of the globe will remain so for centuries [2]. These issues are pointing out at coal gasification as an emerging and alternative process to deliver electricity, synthetic fuels for industrial feedstock and hydrogen for the energy and transportation sectors, thus addressing security of fuel supply and sustainable economic development.

There are several technologies that can be employed in gas separation processes. These include conventional and mature technologies such as cryogenics, solvent extraction, adsorbents in pressure or temperature swing adsorption, and low temperature polymeric membranes. These technologies are of very low industrial risk but have limited scope for additional technological advancement. Further, many of these technologies require low temperatures (<100°C) to operate efficiently. This requirement adds considerable energy penalties in cooling gases from coal gasification. To address this issue, several agencies in the USA (US DoE), Japan (NEDO), Australia (CO2CRC and cLET), and Germany (BMWA/COORETC) are funding R&D in novel materials and high temperature membranes classed as novel and enabling technologies which are currently at an embryonic or growth stage. In this group the major research interest is in material development for membrane technologies including composite high temperature polymers (polyamide and composites), metals and ceramics including molecular sieves (silica and carbon), or novel materials for PSA application [3]. This work assesses the development of high temperature membranes (ceramics and metals) to address the current and future needs for H_2 separation and syngas processing in coal gasification.

Coal Gasification

An Integrated gasification Combined Cycle (IGCC) scheme for the production of hydrogen with CO_2 capture for bituminous coals using the current best available technology is shown in Figure 1. The process can be divided into four steps as follows: (1) air separation &gasification, (2) gas cleaning, (3) gas conditioning, and (4) gas separation.

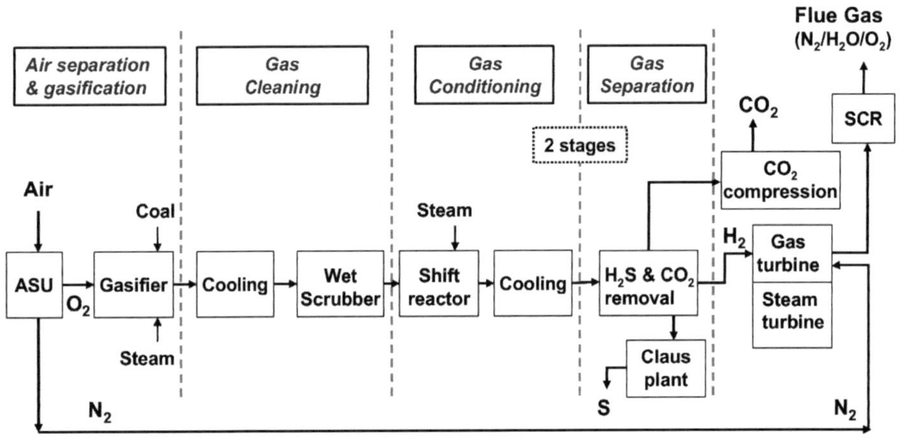

Figure 1 A conventional scheme for IGCC with CCS

In the first step the coal is gasified in an oxygen-blown, entrained flow gasifier to produce the syngas, which is then cooled and scrubbed in the second step to remove particulates. In the third step the syngas is then shifted in sour water gas shift reactors. The water gas shift reaction is exothermic and the conversion is limited by thermodynamic equilibrium; the conversion to H_2 & CO_2 decreasing with increasing temperature. The reaction is therefore carried out in two stages, in high and low temperature shift reactors with interstage cooling, and further cooling to reduce the temperature to the temperature acceptable by the next step. In the fourth step gas separation using a physical solvent (e.g. Selexol or Rectisol) is carried out. A two stage unit is needed, the first to remove H_2S for recovery from a concentrated stream (>50%H_2S) as S in a Claus plant, and the second to remove CO_2 for compression and storage. The hydrogen is then combusted and the heat recovered in a conventional combined cycle to generate electrical power, with the nitrogen from the ASU re-injected to the gas turbine to abate temperature and a selective catalytic reduction (SCR) unit to remove NO_x from the flue gases, which will be mainly N_2, H_2O and O_2.

The scheme shown above is clearly complex and capital intensive, with thermodynamic limitations to the overall efficiency. Although the capability for co-production of hydrogen when demand for electricity is reduced is an asset when there is a local user, there is potential for improvement by combining the gas conditioning and separation steps and the next development in this direction being shown in Figure 2. The cooling before H_2S and CO_2 removal has been eliminated by substituting a warm dry gas cleaning system, comprising a particulate filter and solid sorbent desulphuriser operating at 300 to 500°C, and using a warmer temperature membrane to separate the H_2 from the CO_2. The membrane is integrated with the shift reactor in such a way that the removal of the hydrogen reduces the concentration within the catalyst bed, thereby driving the achievable conversion higher at a specific temperature.

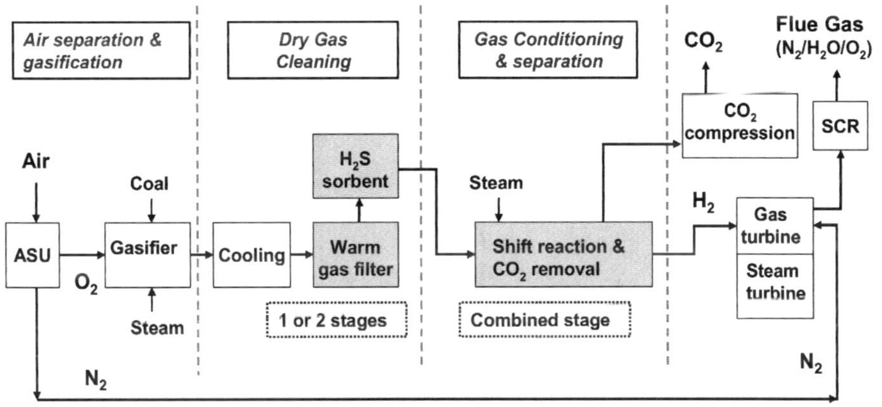

Figure 2 An Improved scheme for IGCC with CCS

A further advance is depicted in Figure 3 below; here a higher temperature membrane allows further simplification of the gas cooling, conditioning and separation step, but this scheme requires both a low cost high temperature membrane tolerant to the syngas and a hot gas cleanup system capable of particulate and sulphur removal at hot conditions between 600 and 800°C

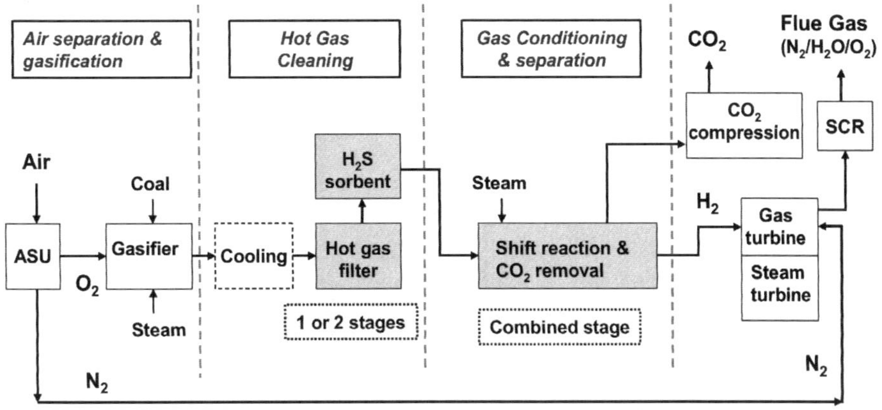

Figure 3 An Advanced scheme for IGCC with CCS

The schemes shown above in Figures 2 & 3 require some development of the units highlighted, as being conducted under the R&D program directed by the Centre for Low Emission Technology (www.clet.net.)

Metal Membranes

Dense metal membranes based on Pd-alloys are in many respects ideal for separating high-purity H_2 from mixed gas streams at high temperatures [4-5]. These membranes operate by dissociating molecular H_2 on the membrane surface and interstitial diffusion of atomic H across the membrane. Recombinative formation of H_2 on the downstream surface completes the transport of H_2 across the membrane. Because this mechanism is only feasible for H_2, the selectivity for metal membranes with low defect densities for H_2 relative to other gases can be very high.

Considerable research has focused on fabricating membranes based on ultra-thin metal films, since reducing the thickness of the film increases H_2 throughput. Most approaches are based on thin metal films on macroporous ceramic or stainless steel supports. Several methods now exist to produce high quality supported membranes with metal films less than 5 μm thick [6]. Recent reports also exist for microfabrication of essentially freestanding thin metal membranes [7].

The two key issues for the development of metal membranes for coal gasification are achieving high H_2 permeation rates and developing materials that are not susceptible to poisoning by trace contaminants, especially sulphur. Another important avenue for enhancing hydrogen permeation through metal membranes is to find alloy compositions that achieve this goal. Experimental studies have examined a large number of binary Pd-alloys for this purpose, but fewer studies exist for ternary alloys. Recent developments in quantum mechanical modeling of H transport through metal alloys seem likely to accelerate efforts to identify promising ternary Pd-alloys [8].

Ceramic Membranes

Porous ceramic (i.e. zeolites, silica, carbon and pillared clays) are generally employed for H_2 separation. The most important parameter in these materials is pore size tailoring typically in the region of 3-6 Å. Great pore control can be attained in carbon molecular sieve (CMS) [9] by pyrolysing polymeric materials such as cellulose acetate, polyaramides and polyimides [10]. However, cost has been a limiting factor in the uptake of CMS membranes by the industry. Pillared clay membranes are tailored using lamellas of nanometer thickness in layered clay as natural blocks suitable to build a framework of nanometer dimension [11]. However, Vercauteren et al. [12] found that clay platelets are oriented mainly parallel to the surface of the membrane support thus resulting in a high degree of tortuosity and resistance to permeation.

The MFI structure (i.e. silicalite or its aluminosilicate analog, ZSM-5) has been most intensively investigated experimentally and theoretically and often employed in laboratory membrane devices [13]. However, MFI zeolite membranes medium-size (~0.55 nm) pore is slightly large for the separation of light gases. To address this requirement, SAPO-34 (dp ~ 0.38nm) zeolite membranes have now been developed to separate H_2/CH_4, H_2/CO_2, and H_2/N_2 binary mixtures [14]. Zeolite membranes are generally fairly robust and stable for industrial applications. Scale up of zeolite membranes has been limited as heat stress during calcination may produce micro cracks, thus rendering the membranes ineffective for gas separation.

Molecular Sieving Silica (MSS) membranes are generally synthesised by a sol-gel process using dip coating [15-18] or CVD methods [19-20], resulting in microporous apertures of molecular dimension from polycondensation of silicon co-polymers. MSS membranes have a high concentration of hydrophilic silanol (Si-OH) groups, making these membranes chemically hydro unstable, which limits the application of these materials to dry feed.

Discussion

Metal and ceramic membranes can operate at high temperature and therefore address the requirements of improved (Fig. 2) and advanced (Fig. 3) IGCC systems. Two of the most important parameters that describe the performance of a membrane are productivity (i.e. permeability) and quality (i.e. selectivity). In turn, trade offs are required in membrane systems to balance permeability and selectivity requirements. MSS membranes are now achieving H_2/CO_2 selectivities > 50, while H_2 permeation is now approaching 10^{-6} $mols.m^{-2}.s^{-1}.Pa^{-1}$, equivalent to 135 litres (STP) of H_2 per square metre of membrane area per minute at 1 atmosphere differential pressure across the membrane. The reported permeabilities of supported palladium membranes vary considerably; a value of $5x10^{-4}$ $m^3.m.m^{-2}.hr^{-1}.bar^{-1/2}$ at 500°C is representative. However, low defect palladium membranes can deliver almost pure H_2 streams at temperatures in excess of 500-600°C, an advantage over the glassy transition temperature of MSS membranes.

With relation to materials, functionalisation for ceramic membranes or alloying for palladium membranes will continue to be the major research focus. In an attempt to resolve the hydro-stability problem in MSS membranes for gas streams containing water vapour, researchers have looked at the removal or substitution of hydroxyl groups. Successful work in this area includes co-valently bonded methyl templates to the silica [21], surfactant templated silica sols [22], enhancing condensation reactions with surfactants [23], and carbonising templates in a silica matrix [24]. The use of metal alloys is arguably more important in developing membranes that are not severely poisoned by S-containing species and other common syngas contaminants. The performance of most Pd-alloys that have been tested for this purpose is significantly degraded by chemical poisoning. Reports do exist, however, of PdCu alloys that resist S poisoning at the elevated temperatures of interest for coal gasification [25]. The chemical mechanism underlying this poisoning resistance is not yet understood and solving this puzzle will likely aid in identifying other robust alloys.

A common challenge for both ceramic and metal membranes is scale up to address the needs of coal gasifier plants. Current costs of ceramic membrane tubes for large operation units are in the order of US$ 1,000 per m^2, or even higher in the case of palladium alloy metals. Large volume gas processing will likewise require a large membrane area (in the km^2 range), resulting in considerable costs in membrane unit operations. Hence, production cost reduction will be the major driver in high temperature membrane technology. Apart from the issues discussed above, robust materials will have to be developed to cope with high temperature, high pressures, sulfur, water and other impurities, in particular if processing dirty syngas. In terms of systems, operational longevity data is very scarce and in situ regenerability of membranes as reported by Duke at al. (2004) may allow for operational cost reduction.

The gaps for technology improvement are therefore very attractive to address technology requirements towards improved (Fig. 2) and advanced (Fig. 3) IGCC systems. There are several thousands of arrays for functionalisation, doping, grafting, alloying and producing mixtures which may provide technological breakthroughs. From a technological view point, there is no single technology at the moment that can address all the future needs of syngas processing. It is likely that several technologies will address specific functions and requirements for gas separation in coal gasification.

Conclusions

The technical advancements in coal gasification processes will require gas separation at high temperatures, ideal for embryonic technologies such as metal and ceramic membranes. Whilst conventional approach to improve fluxes and separation factor has been the research focus, novel approaches in alloying metals or functionalizing ceramic materials are paving the way to designing robust membrane technology. Of particular attention, membranes will have to be able to process and tolerate syngas with trace contaminants, in particular sulphur.

References

1. BP, BP statistical review of world energy. 2002.
2. Williams, R.H. Toward zero emissions for transportation using fossil fuels. in VIII Biennal conference on transportation, energy and environmental policy. **2001**. Monterey, CA:
3. Diniz da Costa,J. C., G. Q. Lu and V. Rudolph, Molecular Sieve Silica (MSS) Membranes in: Encyclopedia of Nanomaterials and Nanotechnology, American Scientific Publishers, California, USA , 5, 723-741, 2004.
4. Sholl, D. S., Y. H. Ma, Dense Metal Membranes for Production of High Purity Hydrogen, MRS Bulletin, **2006** in press
5. Paglieri, S. N. and J. D. Way, Innovations in palladium membrane research, *Sep. Purif. Methods,* **2002**, *31*, 1-169.
6. Tong, J.; Kashima, Y.; Shirai, R.; Suda, H.; Matsumura, Y. Thin defect-free Pd membrane deposited on assymmetric porous stainless steel support, *Ind. Eng. Chem. Res.,* **2005**, *44*, 8025-8032.
7. Tong, H. D.; Gielens, F. C.; Gardeniers, J. G. E.; Jansen, H. V.; van Rijn, C. J. M.; Elwenspoek, M. C.; Nijdam, W. Microfabricated palladium-silver alloy membranes and their application in hydrogen separation, *Ind. Eng. Chem. Res.,* **2004**, *43*, 4182-4187
8. Kamakoti, P.; Sholl, D. S. Towards first principles identification of ternary alloys for hydrogen purification membranes, *J. Membrane Sci.,* **2006**, *in press.*
9. Koresh, J., Sofer, A., Molecular sieve carbon permselective membrane. Part I. Presentation of a new device for gas mixture separation Sep. Sci. and Tech., **1983**, 18,723-734.
10. Jones, C. W., Koros, W. J., Carbon molecular sieve gas separation membranes -I. Preparation and characterisation based on polyimide precursors, Carbon, **1994**, 32 (8), 1419-1425.
11. Kleinfeld, E. R., Ferguson, G. S., Stepwise formation of multilayered nanostructural films from macromolecular precursors, Science, **1994**, 265, 370-373.
12. Vercauteren, S., Luyten, J., Leysen, R., Vansant, E. F., Synthesis and characterisation of a pillared clay membrane. J. Memb. Sci., **1996**, 119, 161-168.
13. Xing, W., Diniz da Costa, J. C., Lu, G. Q., Yan, Z.F., Zeolite membranes for environmental separation and reactions in: Encyclopedia of Nanoscience and Nanotechnogy, **2004**, 1157-1166, Marcel Dekker
14. Poshusta, J. C.; Tuan, V. A.; Falconer, J. L.; Noble, R. D. Synthesis and Permeation Properties of SAPO-34 Tubular Membranes. Industrial & Engineering Chemistry Research, **1998**, 37 (10), 3924-3929.
15. de Vos, R.M., Verweij, H., Improved performance of silica membranes for gas separation. J. Memb. Sci., **1998**, 143, 37-51.
16. Raman, N. K., Brinker C. J., Organic template approach to molecular sieving silica membranes, J. Memb. Sci., **1995**, 105, 273-279.
17. Kusakabe, K., Sakamoto, S., Saie, T., Morooka, S., Pore structure of silica membranes formed by a sol-gel technique using tetraethoxysilane and alkyltriethoxysilanes, Sep. Pur. Tech., **1999**, 16, 139-146.

18. Diniz da Costa, J. C., Lu, G. Q., Rudolph, V., Molecular Sieve Silica (MSS) Membranes in: Encyclopedia of Nanomaterials and Nanotechnology, American Scientific Publishers, California, USA , **2004**, 5, 723-741.
19. Lee, D., Oyama, S.T., Gas permeation characteristics of a hydrogen selective supported silica membrane. J. Membr. Sci., **2002**, **210** (2), 291-306.
20. Suraj G., Ono K., Nomura M., Sugawara T., Nakao S. Microporous silica membrane prepared using TMOS/O3 CVD in opposing reactants geometry, Trans. Mat. Res. Soc. Japan, **2004**, 29, 3263-3266.
21. de Vos, R.M., Maier, W.F., Verweij, H., Hydrophobic silica membranes for gas separation. J. Memb. Sci., **1999**, 158(1-2), 277-288.
22. Tsai, C.-Y., S.-Y. Tam, Y. Lu, and C.J. Brinker, Dual-layer asymmetric microporous silica membranes, J. Memb. Sci., **2000**. 169 (2), 255-268.
23. Giessler, S., Diniz da Costa, J. C., Lu, G. Q., Hydrophobicity of templated silica xerogels for molecular sieving applications, J. Nanoscience & Nanotechnology, **2001**, 1 (3), 331-336.
24. Duke, M., Diniz da Costa, J. C., Lu, G. Q., Grey, P., Petch, M., Carbonised Template Molecular Sieve Silica Membranes In Fuel Processing Systems: permeation, hydrostability and regeneration, J. Memb. Sci., **2004**, 241, 325-333.
25. Kamakoti, P.; Morreale, B. D.; Ciocco, M. V.; Howard, B. H.; Killmeyer, R. P.; Cugini, A. V.; Sholl, D. S. Prediction of Hydrogen Flux through Sulfur-Tolerant Binary Alloy Membranes.*Science* **2005**, *307*, 569-573.

182a Pre-Conceptual Design and Cost Estimate for a Nuclear Hydrogen Production Plant

William A. Summers[1], Maximilian B. Gorensek[2], and Edward T. Danko[1]. (1) Energy Security Department, Savannah River National Laboratory, Aiken, SC 29808, (2) Computational Science and Statistics Department, Savannah River National Laboratory, Aiken, SC 29808
A national energy system based on the use of hydrogen as a primary energy carrier will require new methods for large-scale hydrogen production and distribution. One approach for efficiently producing large quantities of hydrogen without producing greenhouse gases is the use of nuclear water-splitting. The DOE Office of Nuclear Energy, Science and Technology (NE) has established a Nuclear Hydrogen Initiative with the goal to demonstrate the economic, commercial-scale production of hydrogen using nuclear energy. If successful, this research could lead to a large-scale, emission-free, domestic hydrogen production capability to fuel a future hydrogen economy.

This presentation will discuss the results of work performed during a three-year study for the Nuclear Energy Research Initiative (NERI) Program under the auspices of the DOE-NE. The objective of this research was to identify, characterize, and evaluate the critical technical and economic issues associated with centralized nuclear hydrogen production. The technical approach consisted of the thermochemical decomposition of water using heat from a next generation (Gen IV) nuclear reactor. The research team, led by Savannah River National Laboratory, addressed not only the nuclear hydrogen production plant, but also infrastructure issues associated with integrating such a plant into a complete hydrogen supply system. This included consideration of hydrogen production, storage, distribution, and end-user integration, as well as the interaction of these factors.

During the first phase of the project, the design and evaluation was performed for an Nth-of-a-kind commercial nuclear hydrogen production plant. This included identification and analysis of the major requirements for nuclear hydrogen production; definition of the physical characteristics and major considerations for the nuclear reactor, the thermochemical hydrogen production process, and the related hydrogen infrastructure; and an economic analysis of hydrogen production costs, infrastructure costs, end-user economics, commercial hydrogen cost comparison, distributed electrolysis comparison, and tradeoff studies and sensitivity analysis.

During the study phase of the study, a site-specific, pre-conceptual design of a commercial prototype nuclear hydrogen production plant was developed. This test-case design provided for a more detailed evaluation of nuclear hydrogen production, including plant design and site selection issues, integration with an existing industrial hydrogen end-user, sizing and characterization of hydrogen storage and pipeline transmission systems, identification and evaluation of other hydrogen infrastructure issues, and a detailed economic assessment.

182b Experimental Studies of the Iodine-Water System for the Sulfur-Iodine Thermochemical Cycle

Jacob M. Crosthwaite, Chemical & Biomolecular Engineering, Clemson University, 127 Earle Hall, Clemson, SC 29634, David Bruce, Department of Chemical and Biomolecular Engineering, Clemson University, 127 Earle Hall, Clemson, SC 29634-0909, and Mark C. Thies, Clemson University, Dept of Chemical Engineering, Earle Hall, Clemson, SC 29634-0909.
Of the 100+ thermochemical hydrogen cycles that have been proposed, the sulfur-iodine (S-I) cycle is the primary target of international interest for the centralized production of hydrogen from nuclear power. However, this cycle involves complex, highly nonideal phase behavior that is poorly understood, so current performance projections are based on uncertain and incomplete data. Thus, thermodynamic measurements and physical property models for the analysis of the S-I cycle have been identified as a basic research need for the Hydrogen Economy. Only with reliable data, along with carefully constructed property and process models that correctly describe the process, can the true potential of the S-I cycle be assessed.

The current focus of our experimental work is the measurement of phase behavior relevant to the HI decomposition section (i.e., the reactive distillation column) of the S-I cycle, that is, measurements for mixtures of HI, molten iodine, and water at elevated temperatures and pressures. Process simulation of the S-I cycle by our DOE-funded project team has identified the measurement of liquid-liquid equilibrium (LLE) data for the iodine-water system as the highest priority binary for measurement, so initial work is proceeding with this system. Only one study (F. C. Kracek, *J. Phys. Chem.*, **1931**, 417-422) has been reported for the iodine-water binary, and those results extend only up to ~200 °C. Our goal is to not only validate this early work, but to extend measurements up to the mixture critical point for the iodine-water system, which is expected to occur in the vicinity of 300 °C.

A continuous-flow apparatus with a windowed equilibrium cell has been designed and constructed to measure equilibrium phase compositions for HI-I_2-H_2O mixtures at temperatures to 350 °C and pressures to 200 bar. Because of the highly corrosive nature of these mixtures, especially at elevated temperatures, the choice of materials for all wetted parts in the continuous-flow setup was an important issue in terms of ensuring both safety and reliability. Details of the construction of this apparatus, materials selection issues, and initial LLE results for the iodine-water binary are to be presented.

SULFURIC ACID DECOMPOSITION CATALYSTS AND REACTION CONSIDERATIONS FOR SULFUR-BASED THERMOCHEMICAL WATER SPLITTING CYCLES

Daniel M. Ginosar, Harry W. Rollins, Lucia M. Petkovic, and Kyle C. Burch

Idaho National Laboratory
Chemical Sciences
P.O. Box 1625
Idaho Falls, ID 83415-2208

Abstract

The sulfur-based family of thermochemical cycles appears promising for producing hydrogen from water. These cycles employ a high-temperature sulfuric acid decomposition reaction step. The reaction takes place from 700 to 900 °C, or higher, and is facilitated by heterogeneous catalysts. The overall energy efficiency of the hydrogen production process is directly correlated to the temperature of the sulfuric acid reaction step. The activity of spinels and perovskites were explored due to their potential to be stable at the high reaction temperatures. This paper presents the temperature dependent activity, from 725 to 900 °C, of several complex metal oxide catalysts for the decomposition reaction: two perovskites ($FeTiO_3$ and $MnTiO_3$) and four spinels ($NiFe_2O_4$, $CuFe_2O_4$, $NiCr_2O_4$, and $CuCr_2O_4$). For comparison, a 1 wt% Pt/TiO_2 catalyst and two common metal oxides, CuO and Fe_2O_3, were also studied. The most promising material examined, a $CuFe_2O_4$ spinel had significantly higher activity than the 1% Pt/TiO_2 catalyst at temperatures of 850 °C and above, but much lower activity than the supported platinum catalyst at temperatures below 825 °C. These results suggest that to minimize reactor size and maximize energy conversion efficiency for the water splitting cycles, a minimum of two catalyst zones, utilizing both a supported Pt and a spinel catalyst should be employed; a lower temperature section employing a Pt/TiO_2 catalyst followed by a high temperature section utilizing the spinel catalyst.

Introduction

Thermochemical cycles produce hydrogen through a series of chemical reactions where the net result is the production of hydrogen and oxygen from water at much lower temperatures than from direct thermal decomposition. All chemicals within the cycle are fully recycled and the heat to drive the reactions, which tend to be endothermic, must be provided by a primary energy source. When the primary energy driver is nuclear heat, hydrogen can be generated without producing green-house gasses, and can provide independence from our dwindling supplies of fossil fuels.

The sulfur-based family of thermochemical cycles appears to have significant technical viability for producing hydrogen.[1] All of the sulfur-based thermochemical cycles, including the Sulfur-Iodine cycle,[2] the Hybrid Sulfur cycle,[3] and the Sulfur-Bromine Hybrid cycle,[4] incorporate the sulfuric acid decomposition reaction as the high temperature step in their cycles. This reaction decomposes sulfuric acid to oxygen, steam, and sulfur dioxide. The production of oxygen represents half of the water splitting reaction and is a product from the cycle. The generated steam and SO_2 are recycled for use in the other reaction steps to complete the water splitting reaction.

The sulfuric acid decomposition reaction takes place at temperatures between 700 and 900 °C, or higher, and is catalyzed by a heterogeneous metal oxide or precious metal catalyst. This reaction is the highest temperature unit operation in the thermochemical hydrogen production process and dictates the temperature requirements of the primary energy source. Due to thermodynamic limits, high temperatures for the sulfuric acid reaction are necessary to achieve high levels of chemical conversion, and hence energy efficiency.

The efficiency of Sulfur-Iodine cycle is highly dependent on the sulfuric acid decomposition reaction temperature. Preliminary analysis reported by Brown et al, estimated that the efficiency of the Sulfur-Iodine cycle could be as high at 52% at an operating temperature of 900 °C, but declines to 42% at 827 °C, and is less than 30% at 750 °C.[1] This decline is due to the thermodynamic limit of the endothermic sulfuric acid decomposition reaction. At the highest anticipated reaction temperature of 900 °C, acid conversions of up to 88 % are possible, but decline with decreasing temperature to 66% at 750 °C. Significant levels of energy are required to recover and reheat the unconverted acid for recycle to the decomposition reaction. Thus, it is vital to operate this reaction at the highest possible temperature.

Catalysts for sulfuric acid decomposition were initially explored during the 1970s and 1980s.[5-10] Activities of over 30 catalysts have been reported. Those studies found that there were few highly active catalysts. Most metal oxides had low activity due to the formation of sulfates on the catalysts.[6-9] The formation of sulfates was more significant at lower reaction temperatures and was predicted to be problematic as reaction pressure increased to anticipated operating values.[8,9] During that time period, the long-term stability of the few promising catalysts had not been reported in the open literature.

Recently, we reported on the activity and stability of catalysts composed of platinum supported on porous metal oxides.[11,12] In that work we examined catalysts comprised of low weight fractions of platinum supported on γ-Al_2O_3, ZrO_2, and TiO_2 with support surface areas of 180, 51 and 2 m^2/g, respectively. It was found that initial conversions increased with support surface area, but only the low surface area catalyst was stable over short duration testing.[11] Activity and stability was then examined for platinum on low surface area supports.[12] Supports studied were TiO_2 and α-Al_2O_3 which were considered to be stable at high temperatures and acid resistant. However, at 800 and 850 °C, for reaction times of 24 hours, only the Pt/TiO_2 catalysts were stable. Longer term studies for 240 hours at 850 °C showed that even the Pt/TiO_2 was not stable. Catalyst deactivation appeared to be due to permanent damage to the active metal through sintering and volatilization of platinum.

In this work, we discuss the activity of alternative complex metal oxide catalysts: spinels and perovskites. In general, these oxide catalysts have lower activity than platinum catalysts, but appear to have better high-temperature stability. Implications of catalyst activity and stability on reactor design are discussed.

Experimental Methods

A high-temperature continuous flow catalyst test system was used to examine the catalytic reaction in flowing, concentrated sulfuric acid at atmospheric pressure. The system was built of Teflon tubing and glass and employs a high-temperature quartz reactor. The system includes two top loading balances for sulfuric acid feed and collection mass balance, a pulse-free peristaltic feed pump, a mass flow controlled reference gas stream, and a gas chromatograph (GC) for product quantitation.

In the test system, concentrated sulfuric is pumped into a heated zone where the acid vaporizes and decomposes to SO_3 and steam. Generated vapors flow over a bed of catalyst for the reduction of SO_3 to produce SO_2 and oxygen. Any unreacted SO_3 exiting the reactor recombines with H_2O to form sulfuric acid which is condensed and collected. The remaining gasses combine with a reference gas stream which together flow through a GC sample valve and are then vented to a hood. Product concentrations of SO_2 and O_2 are determined approximately every 20 minutes by GC analysis. Product rates were determined from measured concentrations and known reference gas mass flow rate.

Catalyst testing focused on mixed metal oxides; two perovskites ($FeTiO_3$ and $MnTiO_3$) and four spinels ($NiFe_2O_4$, $CuFe_2O_4$, $NiCr_2O_4$, and $CuCr_2O_4$). For comparison, a 1 wt% Pt/TiO_2 catalyst and two common metal oxides, CuO and Fe_2O_3, were also studied. The metal oxides were all obtained as nano-scale powders. The temperature dependent activities of these materials were measured from 725 to 900 °C at 25° increments. Liquid sulfuric acid at 95 to 96 wt% was pumped into the catalytic reactor at 48 g/hr. One gram of catalyst was loaded into the reactor resulting in a pure acid weight hour space velocity (WHSV) of about 46 g-acid/g-catalyst/hr.

Following reaction, catalyst samples were recovered and analyzed for property degradation. BET surface areas were determined by nitrogen physisorption on an automated Quantachrome Autosorb-1C system. The catalyst samples were submitted to outgassing for 3 h at 300 °C prior to the analyses. B.E.T. surface areas were calculated in the range of P/Po between 0.05 and 0.35 and total pore volume was determined at P/Po = 0.99.

Results

Catalyst Activity

Temperature dependent SO_2 yields over the six mixed metal oxides and the supported platinum catalyst are shown in Figure 1. Over the temperature range, spinel and perovskite activity followed the general trend: $CuCr_2O_4 > CuFe_2O_4 > NiCr_2O_4 \approx NiFe_2O_4 > MnTiO_3 \approx FeTiO_3$. At temperatures less than 825 °C, the 1% Pt/TiO_2 catalyst had higher activity than

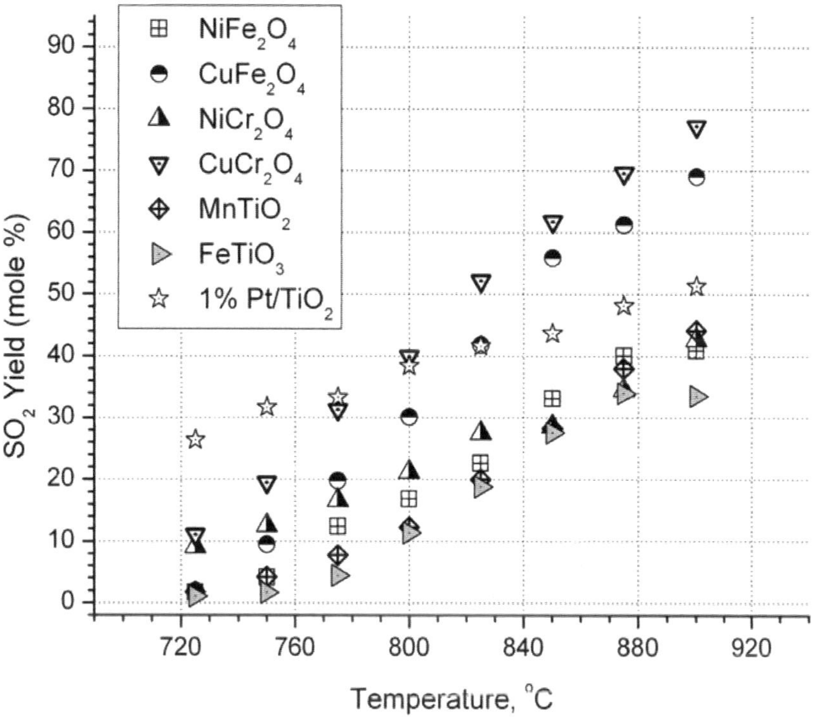

Figure 1. Temperature dependent activity of Spinel, Perovskite and Pt/TiO_2 catalysts based on SO_2 product yield.

most of the metal oxides, but at temperatures above 850 °C, the $CuCr_2O_4$ and $CuFe_2O_4$ spinels had the highest activity.

Although many of these materials displayed promising activity, several also presented potential negative attributes. The $CuCr_2O_4$ and $NiCr_2O_4$ spinels were observed to have leaching problems. The recovered catalyst weights were both much lower than their before reaction weights, the reactors had green discolorations downstream of the catalyst beds, and the recovered liquid acid was tinted with a strong green hue, suggesting that chrome was leached from the catalysts in the reaction environment. The $FeTiO_3$ perovskite also produced a discoloration of the reactor due to minor leaching problems. Further, the $FeTiO_3$ catalyst displayed decrease activity as temperature was increased from 875 to 900 °C, suggesting a lack of high temperature stability. However, $CuFe_2O_4$ catalyst appeared very promising due its high activity and lack of any leaching issues.

To confirm that the activity of the $CuFe_2O_4$ spinel was not due the independent contributions of CuO and Fe_2O_3, its activity was compared to those materials. Figure 2 shows the activity of the spinel and the two metal oxides. The $CuFe_2O_4$ spinel had clearly superior activity to the iron oxide. At temperatures below 875 °C, the spinel had a clear activity advantage over the copper oxide, but similar activity at 875 and 900 °C.

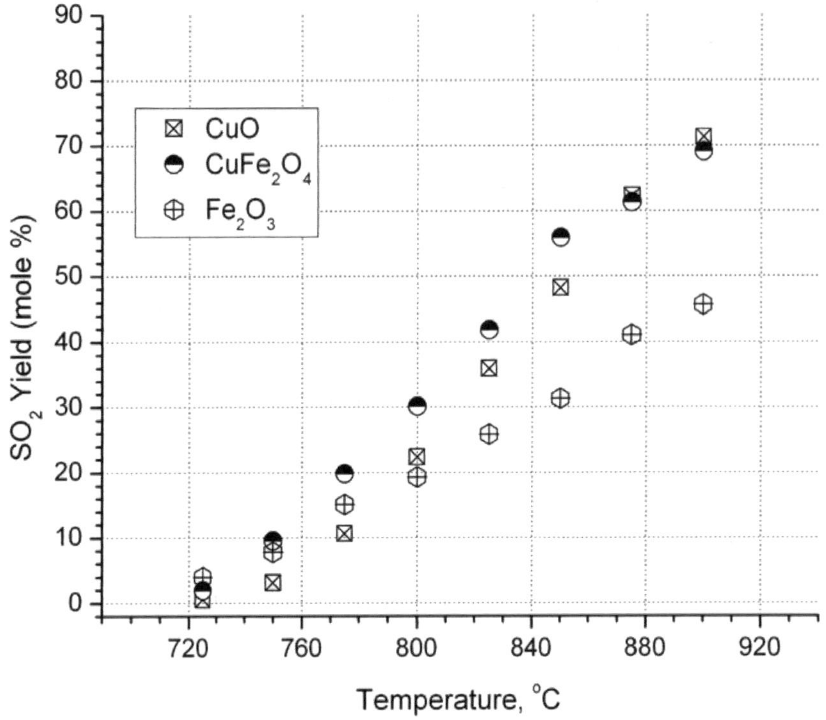

Figure 2. Temperature dependent activity of the $CuFe_2O_4$ spinel and the CuO and Fe_2O_3 catalysts based on SO_2 product yield.

Catalyst Characterization

BET surface area was found to decrease for all of the metal oxides after exposure to reaction conditions. Fresh catalyst surface areas for the metal oxides range from 22 to 97 m^2/g and decreased after reaction to values that ranged from 1.3 to 6.3 m^2/g. Both before and after reaction, the spinel and perovskite materials did not posses measurable micro-pore volume.

Reactor Design Considerations

These results show that the Pt/TiO_2 catalyst is more active than the metal oxides at temperatures below 825 °C, but at temperatures of 850 °C and above, the $CuFe_2O_4$ spinel provided clear advantages. Our previous studies also found that Pt/TiO_2 did not display long term stability at 850 °C and lost activity due to sintering and loss of platinum through volatilization of platinum species.[11] Preliminary calculations suggest that platinum volatilization is reduced by an order-of-magnitude when temperature is decreased by 50 °C. Thus, platinum based catalysts are not appropriate for the higher temperature operations at which the S-I thermochemical cycle is highly efficient, but may be appropriate at temperatures of 800 °C and below. Alternatively, the spinel catalyst has only minor activity at lower temperatures, but displayed significant activity at temperatures of 825 °C and above.

Proper reactor conditions will need to be chosen to optimize reactor design based on catalyst properties. For example, if a maximum operating temperature of 800 °C is chosen in order to take advantage of the higher activity of the Pt catalyst, cycle efficiency would be limited to approximately 38% based on the estimates of Brown et al.[1] The activity of the $CuFe_2O_4$ catalyst between 725 to 800 °C is on average less than half that of the 1% Pt/TiO_2 catalyst, thus, the H_2SO_4 reactor would need to be more than twice as large if the spinel catalyst were used at these temperatures in place of the Pt catalyst. Further, the reaction is highly endothermic, and a large temperature gradient within the reactor is anticipated to be present. The two different catalyst properties along with an anticipated reactor temperature gradient offer the opportunity to minimize reactor size and maximize energy conversion efficiency using a minimum of two reactors, or catalyst zones, utilizing both the Pt based and the spinel catalysts.

A lower temperature reactor employing a Pt/TiO_2 catalyst could be operated up to about 800 °C. This reactor would be followed by a heat exchanger to raise feed temperature such that a high temperature reactor, utilizing the spinel catalyst, could operate in the 850+ °C range. Since the decomposition reaction is highly endothermic, it is essential that heat exchange be incorporated into the catalytic reactor, or multiple reactor/heat exchange stages provided to minimize temperature drop at each reactor stage. Additionally, in order to protect the platinum based catalyst, heat transfer operations should be designed and controlled to limit the maximum temperature of the low temperature reactor. By utilizing this type of strategy reactor size can be minimized and cycle efficiency maximized.

In addition to maximizing conversion efficiency, reactor designs should take into consideration maintenance requirements. Catalysts will deactivate in this harsh reaction environment and will therefore need to be replaced periodically. Depending on deactivation rates, catalyst replacement may need to be accomplished either on-line, or during periodic plant shut downs. In either case, catalysts or catalyst units should be able to be replaced inexpensively and rapidly on-site during short turnaround periods.

Conclusion

The catalytic activities of spinel and perovskite complex metal oxides appear promising for the catalytic decomposition of sulfuric acid. The most promising material examined, a $CuFe_2O_4$ spinel had significantly higher activity than a 1% Pt/TiO_2 catalyst at temperatures of 850 °C and above, but much lower activity than the supported platinum catalyst at temperatures below 825 °C. These results suggest that to minimize reactor size and maximize energy conversion efficiency for the water splitting cycles, a minimum of two catalyst zones, utilizing both a Pt based and the spinel catalysts should be employed; a lower temperature section employing the Pt/TiO_2 catalyst followed by a high temperature section utilizing the spinel catalyst. However, additional study is still required to examine the long term stability of the $CuFe_2O_4$ spinel.

Acknowledgment

This work was supported by the U.S. Department of Energy, Office of Nuclear Energy, Science, and Technology under DOE-NE Idaho Operations Office Contract DE-AC07-05ID14517.

References

1. Brown LC, Besenbruch GE, Lentsch RD, Schultz KR, Funk JF, Pickard PS, Marshall AC, Showalter SK. High efficiency generation of hydrogen fuels using nuclear power. General Atomic Report GA-A24285, June 2003.
2. O'Keefe D, Allen C, Besenbruch G, Brown L, Norman J, Sharp R, McCorkle K. Preliminary results from bench-scale testing of a sulfur-iodine thermochemical water-splitting cycle. Int J Hydrogen Energy 1982;7(5):831-92.
3. Brecher LE, Spewock S, Warde CJ. Westinghouse sulfur cycle for the thermochemical decomposition of water. Int J Hydrogen Energy 1977;2(1):7-15.
4. Beghi GE. A decade of research on thermochemical hydrogen at the joint research center, ISPRA. Int J Hydrogen Energy 1986;11(12):761-71.
5. Spewock S, Brecher LE, Talko F. The thermal catalytic decomposition of sulfur trioxide to sulfur dioxide and oxygen. Proc. 1st World Hydrogen Energy Conf. 1976;1:9A-53.
6. Norman JH, Mysels KJ, O'Keefe DR, Stowell SA, Williamson DG. Chemical studies on the General Atomic sulfur-iodine thermochemical water-splitting cycle. Proc. 2nd World Hydrogen Energy Conf., 1978;2:513.
7. Norman JH, Besenbruch GE, O'Keefe DR. Thermochemical water-splitting for hydrogen production. General Atomic Report GA-A15267, March 1981.
8. Norman JH, Mysels KJ, Sharp R, Williamson, D. Studies of the sulfur-iodine thermochemical water-splitting cycle. Int. J. Hydrogen Energy, 1982;7(7): 545-56.
9. Tagawa H, Endo T. Catalytic decomposition of sulfuric acid using metal oxides as the oxygen generation reaction in thermochemical water splitting process. Int. J. Hydrogen Energy, 1989;14(1): 11-17.
10. High-Pressure Catalytic Metal Reactor in a Simulated Solar Central Receiver. GA Technologies Report GA-A18285, February 1986.
11. Ginosar, DM, Glenn, AW, Petkovic, LM, Burch, KC. Stability of Supported Platinum Sulfuric Acid Decomposition Catalysts for use In Thermochemical Water Splitting Cycles. Int. J. Hydrogen Energy, in press 2006.
12. Ginosar, DM, Petkovic, LM, Burch, KC. Activity and Stability of Sulfuric Acid Decomposition Catalysts for Thermochemical Water Splitting Cycles." Paper 258d, AIChE 2005 National Meeting Overall Conference Proceedings. New York, NY: AIChE. 2005.

Improvement of the Thermal Efficiency of Hydrogen Iodide Concentration in I-S Process by Using Radiation-Induced Polymerized Membrane in Electrodialysis System

Hiroyuki Okuda[1], Tetsuya Yamaki[2] and Shinji Kubo[1],
(1) Nuclear Science and Energy Directorate, Japan Atomic Energy Agency, 4002 Narita-cho, Oarai, Ibaraki, Japan,
(2) Quantum Beam Science Directorate, Japan Atomic Energy Agency, 1233 Watanuki, Takasaki, Japan

Abstract

Thermochemical water-splitting I-S process has been studied as the next generation large-scale hydrogen production method. One of the big issues at stake in this process is to get over HI azeotrope concentration. Up to now, extractive and reactive distillations have both been proposed to deal with this issue. However, these methods increase too much the complexity of the system. Consequently, we have proposed a simple, efficient way of concentrating HI by using electrolysis. Compared to water electrolysis, HI electrolysis can be done using a low voltage (approximately 0.2 V). However, the efficiency of the concentration is decreased by a high permeation rate of H_2O through the membrane. Using a membrane made by the radiation-induced graft polymerization method, which method is also considered for the methanol fuel cell, we expect to improve the overall thermal efficiency of I-S process by 3 %.

1. Introduction

1.1 Hydrogen-heat energy conversion efficiency by I-S process

Japan Atomic Energy Agency (JAEA) has been carrying out R&D on the I-S (Iodine-Sulfur) process that was first proposed by General Atomics Corporation [1]. Figure 1 shows the main reactions of the process and adequate temperature.

The efficiency of the heat to hydrogen energy conversion by the I-S process is discussed. The I-S process is composed of three separate stages, the Bunsen reaction part, that produces the two acids, the sulfuric acid decomposition part, and the hydrogen iodine (HI) decomposition part. Because Bunsen reaction is an exothermic reaction, the quantity of heat consumed for the Bunsen reaction part is ideally nearly 0 kJ/mol. It is reported that the quantity of heat consumed for the sulfuric acid decomposition part can be limited to about 420 kJ per mol of produced hydrogen by introducing the multiple-effect evaporation or vapor recompression method [2].

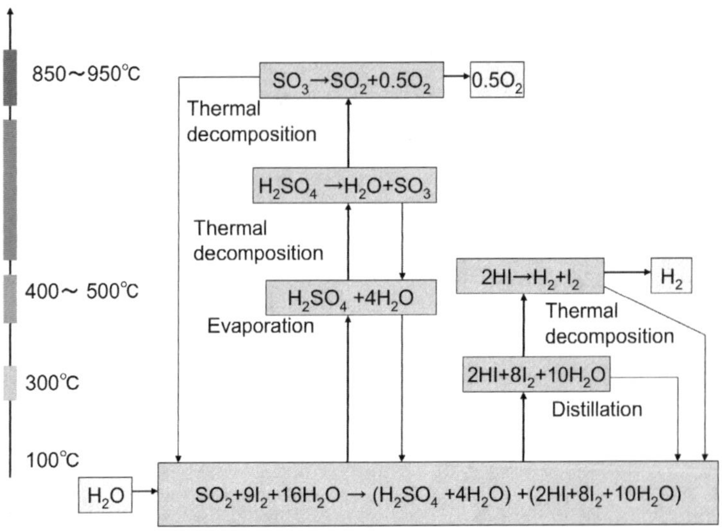

Figure 1. The main reactions of I-S process and adequate temperature.

On the other hand, the quantity of heat consumed for the HI decomposition part is 3380 kJ/mol. The existence of an HI-H_2O azeotrope forbids the separation of aqueous HI by simple distillation. Thus, when evaporating HI, the latent heat of evaporation of the water present in azeotropic HI is also spent and lost.

The efficiency of the heat to hydrogen energy conversion by the I-S process, η, can be written as follows: (the heating value of 1 mole of hydrogen is assumed to be its higher heat value, 286 kJ)

$$\eta = \text{(Heating value of produced hydrogen)} / \text{(Total heat consumed)}$$
$$= (286) / (0 + 420 + 3380) = 7.5\ \%$$

If we hope for an energy conversion efficiency of 40 % (resp. 35 %), the quantity of heat used for HI decomposition must be reduced from 3380 to 295 kJ/mol (resp. 380 kJ/mol).

In other words, in order for the I-S process to reach energetic efficiencies from 35 to 40 %, we must be able to carry on the HI decomposition section using from 300 to 400 kJ/mol of produced H_2, that is we need to decrease the expenditure of this part by a factor of 10 compared to the case of simple distillation.

As means to break azeotrope, extractive distillation and pressure sensitive distillation (PSD) are known. It is reported that the azeotrope is broken by the use of phosphoric acid for the extractive distillation process [1]. PSD (or reactive distillation) is a method to break off azeotrope using the pressure dependence of the azeotropic composition [3]. Both can reduce the heat quantity needed for the separation of HI and H_2O, but new problems occur. In the case of the phosphoric acid extraction distillation, the amount of heat that should be exchanged between solutions including phosphoric acid is quite large and the technical and economical feasibility of the process should be studied.

In order to evaluate reactive distillation, vapor liquid equilibrium data of HI-I_2-H_2O (hereafter termed “HIx solution”) at high pressure condition are necessary. However, because HIx solution is highly corrosive, such data is difficult to acquire experimentally. Therefore VLE is obtained by a thermodynamic calculation, but the precision of data depends on the model of calculation. Those are the basic solutions that will be studied from now on in order to reduce the heat needed in this section by a factor of 10.

In contrast, we suggested an idea to reduce the needed heat for HI concentration by using electrolysis of iodine [4, 5]. In this paper, the possibility of this electrolytic step for I-S process is reported.

1.2 The concept of HI concentration by iodine electrolysis

Figure 2 shows the concept of HI concentration by iodine electrolysis. HIx solutions are separated by a cation exchange membrane. At the cathode, iodine (I_2) is resolved to iodide ion (I^-) while I^- is oxidized to I_2 at the anode. And proton (H^+) also permeates from anode to cathode. As a result, HI component is diluted and I_2 component is concentrated in the HIx solution at the anode, while HI component is concentrated and I_2 component is diluted in the HIx solution at the cathode. HI component of HIx solution is concentrated over the azeotropic concentration by this method, and supplied as feeding liquid of distillation column.

Figure 3 shows the flow chart of the electrolytic HI concentration. The HIx solution, obtained from the Bunsen reaction is divided in two flows by use of the flow separator. HI is concentrated in one of the flows while diluted in the other. The HI rich solution is fed to the distillation column from which high density HI (nearly equal 100 wt%) is obtained at the top and azeotrope HIx solution is obtained at the bottom. It is convenient for the following HI decomposition because HI obtained from the tower top hardly contains any water.

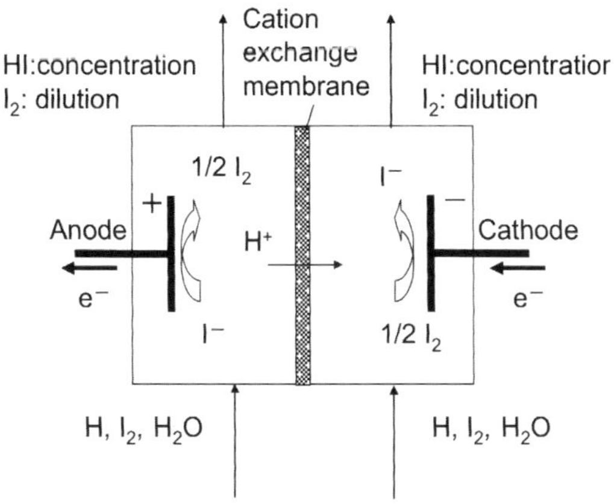

Figure 2. Concept of HI concentration by iodine electrolysis.

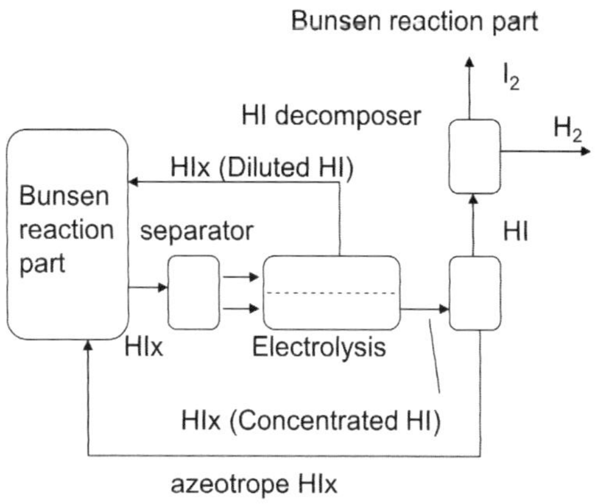

Figure 3. The flow chart applied the electrolytic HI concentration.

1.3 Hydrogen-heat energy conversion efficiency applied to electrolysis

The efficiency of the heat to hydrogen energy conversion applied to electrolysis is discussed. The energy for HI concentration by electrolysis is 70 kJ_e/mol when we suppose it carried out by the voltage of 0.2 V at 100 mA/cm^2. Assuming an electricity-heat energy conversion efficiency of 40 %, a heat of 175 kJ_T/mol of HI is needed (in the case electricity-heat energy conversion efficiency is assumed to be 30%, this figure becomes 233 kJ_T/mol). At this time, from 220 to 280 kJ_T/mol are necessary in the distillation column, if HI is concentrated up to 13 mol/kg_H_2O by electrolysis. (Note: mol/kg means HI 1mol and H_2O 1kg)

1.4 Proton transport number and water permeation coefficient

HI is decomposed in the EED cell where iodine undergoes electrolysis. Two important parameters must be taken in account, together with the transport of H+ through the membrane towards the cathode:

1) I^- moving as a counter ion towards the anode of the EED cell ;
   the proton transport number accounting for this is

$$t = H^+ / (H^+ + I^-)$$

2) Water permeating through the membrane together with H^+, with the water permeation coefficient

$$\beta = H_2O / H^+$$

t and β are important parameters in the evaluation of the efficiency of electrolysis for HI concentration. If β is about 5 for example, the EED cell doesn't achieve its purpose, as the H_2O: H^+ ratio is 5 at azeotropic composition. Figure 4 shows the principle of the electrolytic concentration when t and β are considered.

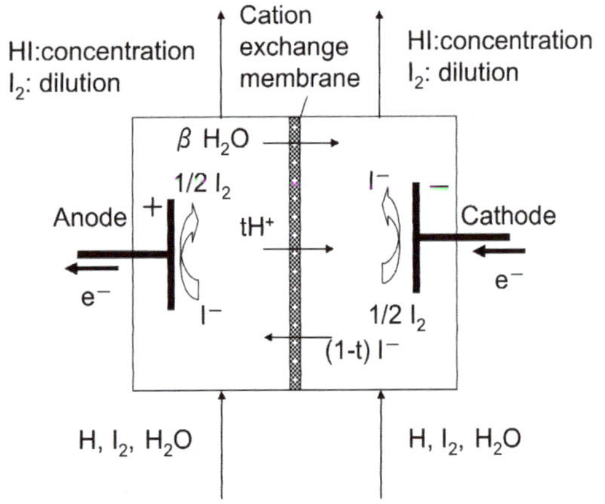

Figure 4. The principle of the electrolytic concentration when t and β are considered.

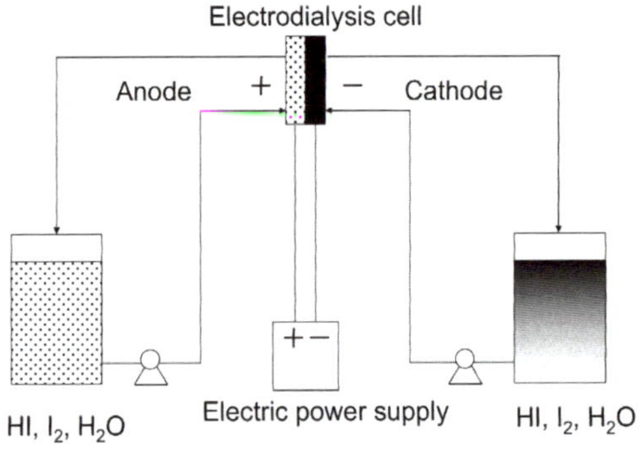

Figure 5. Electrolysis test system of HI concentration using the cation exchange membrane.

2. Experimental

2-1 HI concentration of HIx solution by iodine electrolysis

Figure 5 shows the experimental setup of HI concentration by electrolysis. The quantity of HIx solution fed to the anolyte- and the catholyte-reservoir were HI/ I_2/ H_2O = 114.7/ 224.7/ 90.1 g and 109.8/ 215.2/ 86.3 g, respectively. Effective membrane size was 22 × 22 mm (5.0 cm^2). Figure 6 shows the molality of HI and I_2 at the cathode and the anode for an electrolysis performed at the cell temperature 40 °C, using a current density supply of 200 mA/cm^2 and Nafion117 membrane. The result shows HI concentration is possible; however, it is confirmed that the HI concentration rate decreases after the third hour.

2-2 Membrane property and concentration efficiency

The HI concentration experiment was performed by using the cation exchange membrane produced with the radiation-induced graft polymerization method. The graft polymerized membrane has the following features.

- The water permeation coefficient β has been reported to be small in the case of to the use for the methanol fuel cell. In this lies the advantages of the graft-polymerized membrane compared to Nafion.
- Because the graft rate can be easily changed, various characteristics of the membrane like the proton conductivity can be changed. On the other hand, Nafion offered commercially has constant proton conductivity in general.
- Compared to the cost of chemically produced Nafion, the cost of radiation-induced graft polymerized membranes can be further reduced by simplification and scaling-up of the production process.

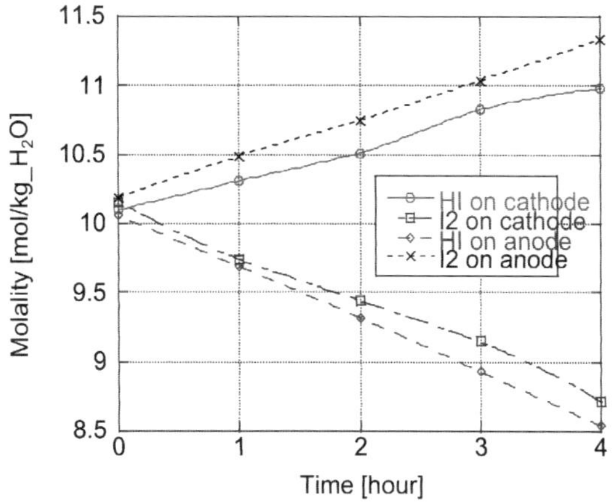

Figure 6. The molality of HI and I_2 at the cathode and the anode for an electrolysis performed at the cell temperature 40 °C, using a current density supply of 200 mA/cm^2 and Nafion117 membrane.

The experimental results and discussion with radiation-induced graft polymerized membrane will be reported at the Conference.

References

[1] J.H. Norman et al., Thermochemical Water-Splitting Cycle, Bench-scale Investigations and Process Engineering, GA-A16713, 1981.

[2] S. Goldstein et al., Upper Bound and Best Estimate of the Efficiency of the Iodine Sulfur Cycle, Int. J. Hydrogen Energy, vol30, No.6, (2005) 619-626

[3] M. Roth and K. F. Knoche, Thermochemical Water-Splitting though Direct HI Decomposition from H2O/HI/I2 Solutions" Int. J. Hydrogen Energy, vol14, No.8, (1989) 545-549.

[4] K. Onuki et al., J. Membr Sci. 192 (2001) 193-199.

[5] G.-J. Hwang et al., J. Membr Sci. 220 (2003) 129-136.

[6] T. Yamaki et al., Radiation Grafting of Styrene into Cross-linked PTFE Films and Subsequent Sulfonation for Fuel Cell Applications, Rad. Phys. and Chemistry, 67 (2003) 403-407.

182e Development of Separation Processes Based on Membrane Technology for the S-I Thermochemical Cycle

Giampaolo Caputo, Claudio Felici, Alberto Giaconia, and Salvatore Sau. ENEA, via Anguillarese, 301, Rome, 00060, Italy

ENEA (the Italian National Agency for New Technology, Energy and the Environment) is involved in an National R&D 3-years program on thermochemical water-splitting cycles to be coupled with solar-thermal power plants. This project is being carried out in the framework of a integrated national program dealing with the sustainable hydrogen production and storage (TEPSI project). A preliminary screening of the most feasible thermochemical processes for hydrogen generation to be coupled with solar concentration plant has been achieved. On the basis of the literature feasibility information, estimation of the overall theoretical loop efficiency, and the maximum temperature, two thermochemical processes were chosen: sulfur-iodine and mixed ferrites cycles. Both processes are currently under investigation to demonstrate the scientific feasibility at laboratory/bench scale. Intensive theoretical and experimental work on the sulfur-iodine cycle is currently aimed to the assessment of the possible options to carry out the operations within the three main sections of the process loop (Bunsen, HI concentration/decomposition, H2SO4 concentration/decomposition). Hence, membrane technology may be used for the following applications: ion exchange membranes for a Bunsen electrochemical reactor, selective membranes for pervaporation operations, gaseous HI or sulfuric acid decomposition reactors, and, finally, HI concentration by electro-electrodialysis. Here we present our experimental results with assessment of membrane distillation routes to improve HI concentration. This route is simple, with low energetic consumption and does not imply the utilization of special materials thanks to the relatively low operating temperature and pressure. HI concentration from water solutions by membrane distillation was examined with a recirculation batch apparatus. Two commercial hydrophobic membranes have been used with two different configurations: direct contact membrane distillation (DCMD) with a polypropylene (PP) capillary membrane in the first case, and an air-gap membrane distillation (AGMD) with polytetrafluoroethylene (PTFE) flat-sheet membrane in the second case. Operative temperature was 60°C for DCMD and 80°C for AGMD; in both cases the operating pressure was about 1 atm. Results showed that the HI concentration in the feed increased from 0.28 up to 7.0 mol/L for DCMD, and from 0.33 up to 8.0 mol/L for AGMD. This latter value is greater then the azeotropic concentration (7.57 mol/L) of the H2O/HI mixture and, consequently, HI further separation from water can be achieved by means of conventional distillation units. It was determined that the higher the HI concentration in the feed mixture, the higher the HI permeation through the membrane: in fact, from 1 L of feed at 7.2 mol/L HI, 0.7 L feed at 8.0 mol/L HI and 0.3 L of permeate at 5.2 mol/L HI concentration were obtained, with an average flux of 7.81 L/h/m2. In the case of the DCMD process, the maximum feed concentration reached was 7 mol/L, i.e. lower than the azeotropic value, and together with permeate fluxes as low as 0.1 L/h /m2. Durability tests of the membrane were also carried out by soaking the membrane in an aqueous at 57%w/w solution HI at 100 °C. In the first tests only H2O/HI solutions have been examined. In the near future the effect of the presence of residual iodine in the feed mixture (used in excess in the Bunsen section in order to optimize phase separation and minimize side reaction occurrence) will be investigated. New membranes and membrane configurations (e.g. vacuum membrane distillation, VMD) are under investigation as well as the energetic process evaluation even at the higher scales. Results obtained will be discussed and compared with the more conventional methods for concentration and purification of the liquid phases produced in the Bunsen reaction.

182f Technical Barriers and Opportunities in Nuclear Plant/Hydrogen Plant Connection Technologies

Steven R. Sherman, Idaho National Laboratory, P.O. Box 1625, 2525 North Fremont Avenue, Idaho Falls, ID 83415-3865
Nuclear powered hydrogen production can be performed today using liquid water electrolysis. Unfortunately, the cost of hydrogen produced from liquid water electrolysis is not yet economically competitive with hydrogen produced from fossil fuel sources for large-scale applications, and other methods are sought for producing hydrogen that take advantage of a nuclear power source (no greenhouse gas emissions, domestic fuel source, high power density) at lower cost.

Nearly all nuclear-powered hydrogen production processes currently under study – the Sulfur-Iodine process, Calcium-Bromine process, and others – require the use of thermal energy in addition to electricity for splitting water. This necessitates the development of a thermal transmission network to connect the nuclear plant with the hydrogen production plant.

While progress has been made on the development of individual enabling technologies, such as the identification of candidate high temperature materials and heat exchanger designs, and preliminary system modeling, much still remains to be solved in order to design and construct a thermal transmission interface capable of transmitting many megawatts of thermal power over long distances safely, efficiently, and that can operate for sustained periods of time at temperatures up to 1000 ºC.

Within the materials area, corrosion, creep, and the reliability of metal/ceramic bonding techniques remain significant problems. In component design, innovations are needed to develop efficient and economic high-temperature heat exchangers that can resist high mechanical and thermal stresses. In the area of safety, more work is needed to incorporate existing chemical plant operating experience into conceptual designs and on-going safety analyses in order to support future environmental and nuclear licensing documentation.

This presentation will focus on specific materials, component design, and safety issues related to the development of the nuclear plant/hydrogen plant thermal interface. Multiple solutions are possible, but solutions must be balanced with costs. Nuclear hydrogen production will not become a practical reality at an industrial scale if the cost of the hydrogen produced is excessive. The thermal interface must be functional without its cost becoming a deciding factor in the implementation of nuclear-powered hydrogen production technologies.

182g Hybrid Sulfur Cycle Flowsheets for Hydrogen Production from Nuclear Energy

Maximilian B. Gorensek, Computational Science and Statistics Department, Savannah River National Laboratory, Aiken, SC 29808, William A. Summers, Energy Security Department, Savannah River National Laboratory, Aiken, SC 29808, and John W. Weidner, Chemical Engineering Department, University of South Carolina, 301, S Main Street, Swearingen Engineering Center, Columbia, SC 29208.
Conceptual designs for two versions of a Hybrid Sulfur (HyS) process for the centralized production of hydrogen (H2) using an advanced nuclear reactor heat source have been developed and economic analyses carried out. Both flowsheets are based on Proton Exchange Membrane (PEM) electrolyzer technology. One electrolyzer uses sulfur dioxide (SO2) dissolved in concentrated sulfuric acid as the anolyte and has an H2 vapor cathode. In this device, SO2 is oxidized with water at the anode to produce H2SO4, while the protons also formed are transported across the PEM and reduced at the cathode to H2. The other electrolyzer has a pure water catholyte and an anhydrous SO2 vapor anode. Here, SO2 is oxidized with water that diffuses across the membrane into the porous anode to make H2SO4. Concentrated sulfuric acid forms at the anode as a liquid product, while protons are transported across the PEM to make H2 at the cathode that can be easily separated from the water catholyte. The flowsheets were modeled, using Aspen Plus™, and material and energy balances were prepared from the simulation results. The presentation will describe these flowsheets, including unit operations and operating conditions, provide thermal efficiency estimates, and give an estimate of H2 production costs.

182h Solar Production of Hydrogen Using a Cadmium Based Thermochemical Cycle

Lloyd C. Brown, Barry W. McQuillan, and Gottfried E. Besenbruch. General Atomics, 3550 General Atomics Ct., San Diego, CA 92121

While screening thermochemical cycles we determined that the cadmium carbonate cycle has the potential of producing hydrogen at very high efficiency using high temperature solar thermal energy. The cycle consists of three chemical reactions:

$Cd(s) + CO_2(g) + H_2O(l) \Rightarrow CdCO_3(s)$ $T < 300^{\circ}C$,

$CdCO_3(s) \Rightarrow CdO(s) + CO_2(g)$ $T \sim 500^{\circ}C$,

$CdO(s) \Rightarrow Cd(g) + \frac{1}{2} O_2(g)$ $T > 1400^{\circ}C$.

The results are presented on an experimental program aimed at accelerating the rate of the hydrogen production reaction. A flowsheet is presented for the overall process and the cost of hydrogen is estimated.

221a Nuclear Hydrogen Initiative Calcium-Bromine Cycle: Determine the Feasibility of Using Cold Plasma Dissociation for H_2 Generation, and Molten-Spray Contactor for Ca•Br_2 Regeneration

Richard D. Doctor, Mike Henry, C.B. Panchal, Yu Po Lin, Steven A. Lottes, and Robert W. Lyczkowski. Energy Systems Division, Argonne National Laboratory, 9700 South Cass Avenue, Argonne, IL 60439-4815

The Calcium-Bromine (Ca-Br) cycle being investigated is a hybrid cycle for hydrogen production employing both heat and electricity. It is particularly attractive because nearly one-half the required thermodynamic energy for water-splitting is delivered as Gen-4 heat at around 1025K (725°C) and it is envisioned that this temperature will facilitate the engineering of materials when compared to other higher temperature cycles. The work during the current fiscal year reported here focuses on two special aspects of the Ca-Br cycle:

- Determine the feasibility of using cold plasma dissociation for H_2 generation from HBr (consideration is being given in the broadest sense to alternate electrical routes that will produce H_2 from HBr).

- Determine the feasibility of using a molten-spray contactor for $CaBr_2$ regeneration in a continuous mode recognizing that there is a eutectic phase with CaO

The thermodynamic basis for a three-stage Ca-Br water-splitting cycle builds upon pioneering work done on the 4-stage University of Tokyo UT-3 process, but Ca-Br employs a plasma-chemical stage for the recovery of HBr as H_2 and Br_2 as a substitute for the final two stages of UT-3. Earlier process design studies on the 3-stage Ca-Br cycle investigated operation in a semi-continuous mode and found that the practical efficiency was 39 - 45%. This current approach is an advance from earlier process design studies, because a continuous rather than semi-continuous mode will be tested in the laboratory. In the earlier semi-continuous mode, the mass and heat balance includes carrying the thermal load of the Calcium titanate used to support the calcium reagent through the temperature cycling between 1025 K and 850 K, thereby creating inefficiencies. Additonally, as batch vessels shift in their chemical environments between reducing and oxidizing atmospheres there are attendant complications with trace gases adsorbed in the system. Of critical importance is that the work reported here focuses on two special aspects of the Ca-Br cycle that will move the system from a semi-batch mode to continuous operations eliminating these issues.

221b Testing of the Palladium Membrane Reactor and Potential Applications for Hydrogen Production

William Rhodes and Dave Howard. Savannah River National Laboratory, Hydrogen Technology Section, Aiken, SC 29803

A Palladium Membrane Reactor (PMR) was tested at the Savannah River Site (SRS), as part of a collaborative effort between the SRS and Los Alamos National Laboratory (LANL). The PMR was developed at LANL as a simple and effective means for recovering hydrogen isotopes from fusion fuel impurities such as methane and water. PMR technology is also applicable to the reforming of JP-8 and diesel fuel, yielding sulfur-free hydrogen. This latter technology is actively being sought for many applications, such as a 10-kW fuel cell Auxiliary Power Unit (APU) powered by logistic fuel. Past accomplishments and future directions shall be discussed.

221c Logistics Fuel Catalytic Cracking for Hydrogen Generation

Mukund Karanjikar, James Stevens, and Lixin You. Chevron Technology Ventures LLC, 3901 Briarpark, Hosuton, TX 77042

Fuel cells offer a unique way to convert chemical energy of a fuel directly into electrical energy. Thus fuel cell technology is a more efficient power generation method. Proton-Exchange-Membrane fuel cells (PEMFC) in the near future and solid oxide fuel cells (SOFC) in the later future are projected to be the most promising technologies. Since PEMFC require hydrogen, reforming of hydrocarbon based fuels (also known as Fuel Processor) is one of the better known techniques for hydrogen generation.Hydrocarbon fuels such as JP 8 (Jet Propulsion Fuel) and diesel have logistics and safety advantages for military applications compared to compressed hydrogen. Energy densities of both diesel and JP8 (11.8 and 11.1 kWh/Kg respectively) are much higher compared to lower hydrocarbons (e.g. natural gas). Processing of heavy hydrocarbon fuels is particularly challenging due to high molecular weight, sulfur content, and aromatic contents of the fuel. This paper presents simulation as well as experimental results from catalytic cracking of heavy hydrocarbon fuels viz. JP 8 and diesel. The simulations are performed using ASPEN® simulation software. System modeling and overall system efficiency calculations are discussed as part of the simulation effort. The experimental part focuses on parametric analysis of the process, mainly effect of temperature and liquid hourly space velocity on the performance of catalysts. The experiments have been carried out in a quartz reactor (1.2 cm dia) using commercial cracking catalysts. Gas analysis is done using HP 5890 ® GC (hydrocarbons) and Del Mar® sulfur analyzer 5504 (H2S and COS).

221d Dynamic Simulation of Reforming and Regeneration Processes for Fixed-Bed Adsorption-Enhanced Natural Gas Reformer

Lixin You, Balaji Krishnamurthy, Yunquan Liu, Farshad Bavarian, Curtis Krause, and James Stevens. Chevron Technology Ventures LLC, 3901 Briarpark, Hosuton, TX 77042
In this paper, mathematical models are presented for all processes involved in absorption-enhanced-reforming (AER) of natural gas. The dynamic model includes mass balance, energy balance for both gas phase and solid beds, and heat and mass transfer between gas and solid. Empirical correlations are used to describe SMR, shift, CO2 adsorption and CaCO3 calcination kinetics. The model has been successful in predicting the reforming, regeneration and cooling processes. Detailed model results will be presented in the paper.

Conventional Steam Methane Reforming (SMR) is the most common hydrogen production technology. Because the SMR reaction is highly endothermic, conventional SMR requires high temperatures and robust heat exchangers to increase the methane conversion to an acceptable level. This results in high capital and operating costs.

It has been proposed that AER can significantly improve steam reforming process. The CO2 absorption drives the steam reforming and water-gas-shift equilibrium to achieve higher CH4 conversion at lower temperatures. Instead of conventional multi-step SMR process (including reforming and water gas shift reactors), AER needs only a single-step to achieve high H2 purity (about 95% compared to about 75% for conventional SMR). Hence, it is anticipated that this technology can be used for numerous hydrogen production applications such as fuel cell power plants, internal combustion engines, refinery and other industrial applications.

In AER process, CO2 adsorption is exothermic and provides the heat for endothermic steam reforming reaction. The combined reforming and CO2 adsorption is nearly adiabatic, which can lead to a very simple reactor design. After reforming, the CO2 adsorbents have to be regenerated to be used for next reforming cycle. Also, as an intermediate step, the CO2 adsorbent bed needs to be cooled to an optimum temperature for efficient CO2 adsorption during reforming. Therefore, the AER process involves cycles of reforming, regeneration and an intermediate cooling step. Since AER process is inherently dynamic, dynamic simulation is necessary for optimizing the reforming, regeneration and cooling steps.

Acknowledgement: This project which is partially funded by DOE under contract DE-FC36- 03GO13102 is acknowledged.

221e Hydrogen Generation Using Fixed-Bed Adsorption-Enhanced Natural Gas Reformer

Balaji Krishnamurthy, Chevron Technical Venture LLC,, 3901 Briarpark, houston, TX 77042, Kerry K. Spilker, Energy Technology Company, Chevron, 3901 Briarpark, Houston, TX 77024, and Lixin You, Chevron Technology Ventures LLC, 3901 Briarpark, Hosuton, TX 77042.

Absorption enhanced reforming (AER) process is a steam reforming process carried out in the presence of CO2 absorbers. The CO2 absorption drives the steam reforming and water-gas-shift equilibrium to achieve higher CH4 conversion at lower temperatures. Instead of conventional multi-step SMR process (including reforming and water gas shift reactors), AER needs only a single-step to achieve high H2 purity (about 95% compared to about 75% for conventional SMR). Hence, it is anticipated that this technology can be used for numerous hydrogen production applications such as fuel cell power plants, internal combustion engines, refinery and other industrial applications. In a fixed bed approach, the CO2 absorbers are utilized during reforming and need to be regenerated in a fixed interval. CO2 adsorption is exothermic and provides the heat for endothermic steam reforming reaction. Regeneration is an endothermic process and requires heat to release CO2 from the material. Also the process requires an intermediate cooling step following regeneration to efficiently absorb CO2 during reforming. Therefore, the AER process involves cycles of reforming, regeneration and an intermediate cooling step. In this paper, general operation of this cyclic process is presented with an emphasis on various methods of regenerating the bed. Experimental results are also provided along with reactor construction details. Details are also presented on the effect of regeneration process on overall system efficiency.
Acknowledgement: This project which is partially funded by DOE under contract DE-FC36- 03GO13102 is acknowledged.

Analysis of a Novel Process with Integrated CO_2 Capture Using Coal and Thermal Energy from a High-Temperature Nuclear Reactor for the Production of Hydrogen and Electric Power

George Tsatsaronis, Kerstin Kapanke, Ana Maria Blanco Marigorta
Technische Universität Berlin
Institute for Energy Engineering
Marchstrasse 18, 10587 Berlin
Germany

ABSTRACT

A novel process generating hydrogen and electric power with very high efficiency and integrated CO_2 capture has been developed. Thermal energy from a high-temperature (about 900°C) nuclear reactor (HTNR) and a solid carbonaceous fuel (coal) are used as input energy to the process. The process is a true "zero emission process" because (a) no NO_X is formed during coal combustion with sulfuric acid, and (b) the combustion products SO_2 and CO_2 are removed separately as compressed liquids from the overall process. A detailed description of the process is presented. The thermodynamic modeling of the process was carried out using the commercial process simulator Aspen Plus. Hydrogen and electric energy are generated by the process with an overall efficiency that is significantly higher than the efficiencies of competing processes generating either one of the products using solid fuels or heat from a HTNR. In addition to the results of the thermodynamic analysis, a brief overview of the economic feasibility of the process is shown including estimates for the product costs. To understand the cost-formation process in the system, an exergoeconomic analysis has been carried out. The main results are also discussed here.

1 INTRODUCTION

Increasing demand for energy, combined with diminishing fossil fuel resources and concerns about greenhouse gas emissions, has increased the interest in the efficient and cost-effective generation and use of hydrogen. In parallel, research is focusing on the analysis and development of fossil-fuel-fired "zero-emission" power plants. Here, a novel "zero-emission" process for the generation of electric energy and hydrogen using coal and high-temperature heat is discussed.

The initial idea for this concept was developed at the Technical University of Aachen in 1980 [1]. Several studies were carried out to evaluate the so called HECAP (Hydrogen and Electric Power from Carbon-Containing Substances and Process Heat) process [2] - [4]. In the present study, some important changes to the original flow diagram and to the main process parameters (see [5]) were made. In the following, a process description and the main results from the thermodynamic, economic and exergoeconomic analyses of the revised process design are presented.

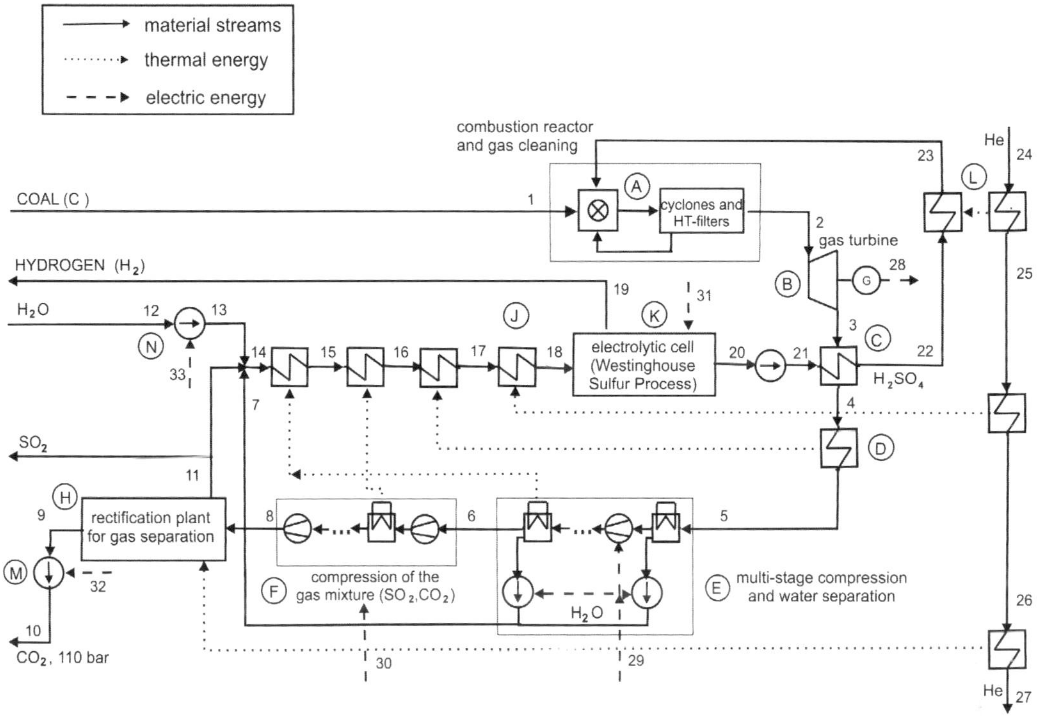

Figure 1. Flow diagram of the HECAP process.

2 DESCRIPTION OF THE PROCESS

The thermodynamic cycle is based on two chemical reactions. The first reaction takes place in an electrolytic cell and delivers the hydrogen product:

$$2SO_2 + 4H_2O \; -> \; 2H_2SO_4 + 2H_2. \tag{1}$$

Sulfur dioxide dissolved in an aqueous solution of sulfuric acid is electrochemically oxidized to sulfuric acid at the anode, while hydrogen gas is evolved at the cathode. In the second reaction, coal reacts with sulfuric acid in a high-pressure combustion reactor to form carbon dioxide, water steam and sulfur dioxide. Using for simplicity pure carbon instead of coal, the chemical reaction is:

$$C + 2H_2SO_4 \; -> \; CO_2 + 2H_2O + 2SO_2. \tag{2}$$

Combining the two reactions, the net reaction of the overall process becomes:

$$C + 2H_2O \; -> \; CO_2 + 2H_2. \tag{3}$$

Carbon reacts with water to form carbon dioxide and hydrogen. Sulfuric acid and sulfur dioxide are involved in the process as recycled intermediate substances. In addition to coal, the input of thermal energy to the overall process is needed because the resulting net reaction (3) is endothermic.

A flow diagram of the proposed thermodynamic cycle is shown in Figure 1. The electrolyzer (plant component K) corresponds to the electrolytic unit of the Westinghouse Sulfur Process for the production of hydrogen [6] to [9], a process that was first proposed in 1973. Hydrogen gas (stream 19) is generated in the electrolytic unit and released as product from the process, while the sulfuric acid solution (stream 20) is used as oxidizer for coal combustion. The sulfuric acid concentration was set to 70 weight-% in the base design. In plant component L, the sulfuric acid solution is vaporized and partially decomposed. Here, the thermal energy demand is covered by a nuclear high-temperature heat production plant. A helium cycle is used to transfer the high-temperature heat from the nuclear plant to the process. From the chemical reaction of coal with sulfuric acid in the combustion reactor (plant component A), a gas mixture consisting mainly of sulfur dioxide, carbon dioxide and water steam is received at a high pressure (50 bar, stream 2). The temperature at the gas turbine inlet is 1644 K for a preheating temperature of 1123 K (stream 23, [3]) in the base design. The gas is expanded in the gas turbine to produce electric energy. Separation of the gases is necessary for recycling the SO_2 and H_2O. The main part of the water (stream 7) can be separated from the mixture by a multi-stage compression and cooling process (plant component E). The remaining gas (stream 6) is further compressed and the components CO_2 and SO_2 are separated in a rectification column (plant component H). The liquid CO_2 stream is pumped to 110 bar for further transportation and storage (stream 10). SO_2 with a small content of water is the bottom product from the column. Any SO_2, formed in the combustion reaction from the sulfur contained in the coal fuel, can be taken out of the process here in liquid form. If pure carbon is considered as fuel, the entire amount of SO_2 is recycled to the electrolytic unit. The operation pressure of the electrolyzer is set to 60 bar. To avoid condensation of SO_2, the operation temperature is chosen at 433 K.

The net electric power output of the cycle is given by the gas turbine power reduced by the electric power consumption in the plant components including all compressors, pumps and the electrolytic unit. The hydrogen gas is delivered from the electrolytic unit at a pressure of 60 bar.

3 THERMODYNAMIC ANALYSIS OF THE PROCESS

The software package *Aspen Plus* [10] was used for the thermodynamic simulation of the cycle. A detailed description of the thermodynamic modeling including assumptions and important parameters are given in [5]. The simulation was based on pure carbon supplied as fuel to the combustion reactor. Full conversion was assumed here. The fuel input rate was set to 0,5 kmol/s carbon to produce a mole flow of 1 kmol/s hydrogen. The exergy of the flow streams was calculated according to the definitions given in [11]. The physical exergy was calculated based on the thermodynamic properties calculated by *Aspen Plus*. The values of the standard chemical exergy of chemical components were taken from [12].

Main results of the simulation are shown in Tables 1 and 2 for the base process design with 70 weight-% sulfuric acid solution: in Table 1, the thermodynamic data of all material streams (stream 1 to 27) and electric energy flows (streams 28 to 34) are given including physical ($\dot{E}_j^{PH}$), chemical ($\dot{E}_j^{CH}$), and total exergies ($\dot{E}_j^{TOT}$) of the material streams. Stream 34 is the net electric energy output of the process. Important energetic and exergetic variables of the plant components and the overall process are shown in Table 2. The values of electric power $\dot{W}_k$, heat transfer rate $\dot{Q}_k$, exergy destruction rate $\dot{E}_{D,k}$, exergy rates associated with the fuel $\dot{E}_{F,k}$ and the product $\dot{E}_{P,k}$, exergetic efficiency ε_k and exergy destruction ratio y_k are given. The exergetic variables were calculated considering physical and chemical exergies separately [13]. Regarding the plant components, the highest exergy destruction ratio is associated with the combustion reactor (plant component A, see Table 2). The exergy destruction of this plant component is caused by the irreversibilities of the chemical reaction and the mixing of streams. Almost 10% of the exergy of the fuel supplied to the overall system (or 24,6% of the exergy of stream 1) is destroyed in the combustion reactor.

Table 1. Thermodynamic and exergoeconomic data of the streams (concentration of the sulfuric acid solution: 70%).

stream no.	$\dot{m}_j$ [kg/s]	T [K]	p [bar]	$\dot{H}_j$ [MW]	$\dot{E}_j^{PH}$ [MW]	$\dot{E}_j^{CH}$ [MW]	$\dot{E}_j^{TOT}$ [MW]	c_j^{PH} [€/GJ]	c_j^{CH} [€/GJ]	c_j^{TOT} [€/GJ]	$\dot{C}_j^{TOT}$ [€/h]
1	6,91	298,2	52,0	-14,4	0,01	202,30	202,31	0,00	2,33	2,33	1699
2	153,39	1643,9	50,0	-1042,2	270,31	265,79	536,09	11,09	20,31	15,67	30233
3	153,39	622,4	0,1	-1279,6	22,69	265,79	288,47	11,09	20,31	19,59	20343
4	153,39	453,2	0,094	-1312,7	6,70	265,79	272,48	11,09	20,31	20,09	19704
5	153,39	360	0,092	-1329,9	1,47	265,79	267,25	11,09	20,31	20,26	19496
6	92,74	303,2	3,2	-527,2	4,38	265,54	269,92	91,80	20,31	21,47	20866
7	60,66	303,3	65,0	-970,9	0,62	0,25	0,87	17,41	20,31	18,24	57
8	92,74	433,2	67,0	-523,3	15,51	265,54	281,05	45,09	20,31	21,68	21936
9	22,01	298,4	65,0	-201,4	4,38	8,09	12,47	64,37	21,61	36,63	1645
10	22,01	310,6	110,0	-201,3	4,52	8,09	12,61	65,06	0,00	23,32	1059
11	70,73	419,6	65,0	-340,7	6,26	259,84	266,1	45,09	21,61	22,16	21233
12	17,11	228,2	1,013	-272,2	0,00	0,03	0,03	0,00	0,00	0,00	0
13	17,11	228,4	65,0	-272,1	0,11	0,03	0,15	56,43	0,00	41,38	22
14	148,50	309,7	65,0	-1583,7	7,23	260,13	267,36	41,39	21,61	22,14	21312
15	148,50	335,9	64,5	-1571,6	8,52	260,13	268,65	52,13	21,61	22,58	21834
16	148,50	347,5	64,5	-1566,2	9,37	260,13	269,50	49,83	21,61	22,59	21916
17	148,50	383,8	64,0	-1548,9	13,01	260,13	273,14	41,90	21,61	22,57	22197
18	148,50	433,2	63,0	-1521,3	21,26	260,13	281,39	31,72	21,61	22,37	22663
19	2,02	433,2	60,0	4,0	10,67	234,69	245,36	84,99	36,85	38,94	34396
20	146,49	433,2	60,0	-1464,2	10,96	105,31	116,27	31,72	36,85	36,36	15221
21	146,49	433,2	60,0	-1464,2	10,96	105,31	116,27	31,72	36,85	36,36	15221
22	146,49	519,5	58,0	-1431,1	23,99	105,31	129,29	28,32	36,85	35,27	16415
23	146,49	1123,2	55,0	-1023,8	163,04	220,44	383,48	13,29	23,19	18,98	26207
24	154,70	1173,2	40,0	703,0	726,50	1078,04	1804,53	8,40	0,00	3,38	21962
25	154,70	662,0	38,5	292,3	444,77	1078,04	1522,80	8,40	0,00	2,45	13446
26	154,70	624,6	38,2	264,6	429,79	1078,04	1507,83	8,40	0,00	2,39	12993
27	154,70	623,5	38,0	263,5	428,71	1078,04	1506,74	8,40	0,00	2,39	12960
28							237,40			19,60	16753
29							23,34			19,60	1647
30							15,07			19,60	1063
31							103,60			19,60	7310
32							0,16			19,60	11
33							0,12			19,60	9
34							95,11			19,60	6711

For the base design, 69% of the total fuel energy is supplied in form of thermal energy; 31% of the total fuel energy is supplied in form of coal. Considering the two energy products, electric energy and hydrogen, 72% of the total energy product is generated in form of hydrogen; 28% of the total energy product is electric energy.

Table 2. Thermodynamic and exergetic variables of the main plant components and the overall process (concentration of the sulfuric acid solution: 70%).

plant component	**no.**	$\dot{W}_k$ [MW]	$\dot{Q}_k$ [MW]	$\dot{E}_{D,k}$ [MW]	$\dot{E}_{F,k}$ [MW]	$\dot{E}_{P,k}$ [MW]	ε_k [%]	y_k [%]
combustion reactor	A	0,00	-3,93	49,66	190,30	140,70	73,9	9,9
gas turbine	B	-237,42	0,00	10,20	247,60	237,40	95,9	2,0
preheater I (H_2SO_4)	C	0,00	33,11	2,96	15,99	13,03	81,5	0,6
preheater II (H_2O, SO_2)	D	0,00	17,24	1,59	5,23	3,64	69,6	0,3
compression, water separation	E	23,34	-179,29	18,52	23,34	4,82	20,7	3,7
compression (SO_2, CO_2)	F	15,07	-5,83	3,09	15,07	11,98	79,5	0,6
rectification plant	H	0,00	-18,83	3,56	6,65	3,09	46,5	0,7
preheater III (H_2O, SO_2)	J	0,00	27,67	6,73	14,98	8,25	55,1	1,3
electrolytic unit	K	103,60	-42,55	23,36	113,60	90,25	79,4	4,7
preheater IV (H_2SO_4)	L	0,00	410,65	27,54	281,70	254,20	90,2	5,5
pump CO_2	M	0,16	0,00	0,02	0,16	0,14	89,1	0,0
pump H_2O	N	0,12	0,00	0,01	0,12	0,11	89,8	0,0
HECAP process		-95,13	439,43	147,20	500,10	340,50	68,1	29,4

For the overall process, the energetic efficiency is

$$\eta_{HHV} = \frac{\dot{W}_{net} + (\dot{n}\overline{HHV})_{18}}{(\dot{H}_{23} - \dot{H}_{26}) + (\dot{m}HHV)_1} = 59,9\%. \tag{4}$$

The exergetic efficiency is defined by

$$\varepsilon = \frac{\dot{W}_{net} + \dot{E}_{18}}{(\dot{E}_{23} - \dot{E}_{26}) + \dot{E}_1} = 68,1\%. \tag{5}$$

The exergy of the product $\dot{E}_P$ from the overall process is equal to the sum of net electric power and exergy flow rate of the hydrogen stream. The exergy of the fuel $\dot{E}_F$ consists of the sum of the difference in the exergy flow rates of the helium stream and the exergy flow rate of the coal (C) stream. Figure 2 shows the effect of the sulfuric acid concentration on the efficiencies of the overall process. For the exergetic efficiency, a maximum value of 68,4% was calculated for 60 weight-% sulfuric acid concentration. The energetic efficiency is continuously increasing with increasing concentration to a maximum of 61,0% calculated for 80 weight-% sulfuric acid concentration. No simulation was carried out for concentrations higher than 80 weight-%. The exergetic efficiency is higher than the corresponding energetic efficiency. This is mainly due to the fact that for the thermal energy input to the process the energy flow rate is significantly higher than the corresponding exergy flow rate.

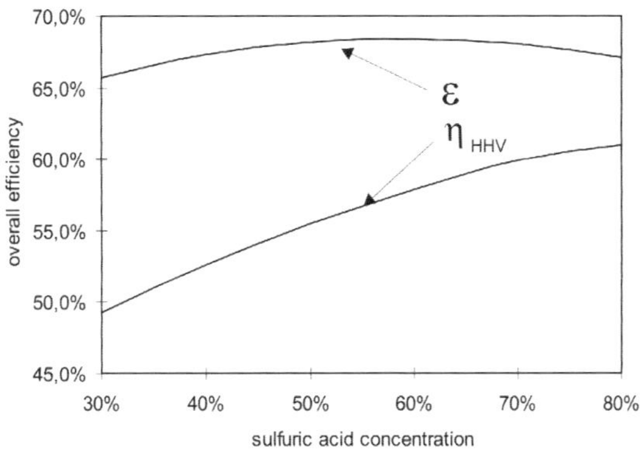

Figure 2. Overall energetic and exergetic efficiency as a function of the sulfuric acid concentration.

4 ECONOMIC AND EXERGOECONOMIC ANALYSES

An economic analysis using the Total Revenue Requirement (TRR) Method [14] was carried out for the proposed design of the HECAP process. In this analysis, levelized values of capital-related charges, fuel costs, and operating and maintenance expenses for the whole plant economic life are calculated. A required return on investment of 10% per year was considered in the analysis. The levelized value of the total revenue requirement of the process are used to estimate the product costs that must be charged for the products of the process. All results from the economic analysis presented here and used in the exergoeconomic analysis are in form of levelized cost values for the whole economic life of the plant (2012-2042). The purchased-equipment costs of every plant component were estimated based on different literature sources [e.g. [15]]. The costs per unit of thermal energy, the time of construction and the plant economic life are based on the assumption that a high-temperature nuclear reactor is delivering the thermal energy. The total annual revenue requirement (TRR) of the project amounts to 295 million €/a (2012-2042). These total cost includes the coal fuel cost (4%), the cost of thermal energy (22%), the operating and maintenance costs (17%) and the carrying charges (57%). The average cost per energy unit of the product for the overall process is 30,74 €/GJ (11,1 c/kWh). The corresponding average costs per exergy unit of the product is 34,40 €/GJ (12,4 c/kWh).

To calculate the costs of each product stream generated in the process, to understand the cost-formation process in the system and to optimize the overall process in future studies, an exergoeconomic (thermoeconomic) analysis was carried out. Combining the results of the exergetic and the economic analyses, the cost rates of all exergy streams within the process were calculated according to the definitions given in [11] and [13]. For the material streams, costs were separately assigned to chemical and physical exergies. The results are presented in Table 1. The costs per exergy unit for physical c_j^{PH} , chemical c_j^{CH} , and total exergy c_j^{TOT}, and the cost rate associated with the total exergy streams $\dot{C}_j^{TOT}$ are given. Based on the results of the economic analysis, the cost per unit of chemical exergy of the coal (stream 1) was set to 2,33 €/GJ. The specific cost of the physical exergy of the helium stream (stream 24) delivering the thermal energy input was set to 8,40 €/GJ. The cost per exergy unit for the incoming water was set to zero. A total cost rate of 1059 €/h calculated for the CO_2 flue gas stream (stream 10) is associated with the physical exergy of the CO_2 leaving the system in compressed form at 110 bar. For the electric energy product (stream 34), the specific cost value from the exergoeconomic analysis is 19,60 €/GJ. The total cost flow rate of the hydrogen product (stream 19) consists mainly of the cost flow rate of the chemical exergy but

Table 3. Exergoeconomic variables of the main plant components and the overall process (concentration of the sulfuric acid solution: 70%).

plant component	**no.**	$\dot{C}_{D,k}$ [€/h]	$c_{F,k}$ [€/GJ]	$c_{P,k}$ [€/GJ]	$\dot{Z}_k$ [€/h]	$\dot{Z}_k + \dot{C}_{D,k}$ [€/h]	f_k [%]	r_k [%]
electrolytic unit	K	1738	20,67	86,48	19644	21382	91,9	318,4
gas turbine	B	407	11,09	19,60	6863	7270	94,4	76,7
combustion reactor	A	417	2,33	7,75	1698	2115	80,3	232,4
preheater IV (H_2SO_4)	L	833	8,40	10,70	1275	2108	60,5	27,4
compression, water separation	E	1307	19,60	112,30	302	1609	18,8	473,2
rectification plant	H	502	39,13	165,90	909	1410	64,4	324,1
preheater I (H_2SO_4)	C	118	11,09	25,45	555	674	82,5	129,4
compression (SO_2, CO_2)	F	218	19,60	26,71	88	306	28,9	36,3
preheater III (H_2O, SO_2)	J	203	8,40	15,68	13	216	5,9	86,7
preheater II (H_2O, SO_2)	D	64	11,09	21,48	73	136	53,3	93,6
pump CO_2	M	1	19,60	63,27	21	22	94,5	222,8
pump H_2O	N	1	19,60	56,43	14	15	93,9	187,9
HECAP process		3150	5,94	34,40	31455	34605	90,9	478,8

around 10% of this cost is associated with the physical exergy of this stream that leaves the systems at a high pressure of 60 bar. The cost per unit of total exergy amounts to 38,94 €/GJ for the hydrogen product. One reason for the relatively high cost of the hydrogen stream is that most costs associated with chemical exergy within the process are assigned to the hydrogen product. Taking into account the cost rate of the flue gas, which can not be sold as a product, the average unit cost per total exergy of the useful product streams for the HECAP process is 34,40 €/GJ.

Exergoeconomic variables calculated for the plant components and the overall process are presented in Table 3; detailed information about the variables used here can be found in [11], [13] and [14]. The cost rate associated with the exergy destruction $\dot{C}_{D,k}$, the average costs of the fuel and the product per exergy unit $c_{F,k}$ and $c_{P,k}$, the cost rate associated with the capital investment and the operation and maintenance costs $\dot{Z}_k$, the exergoeconomic factor f_k and the relative cost difference r_k are given. The plant components are ordered in Table 3 according to their cost importance using the sum $(\dot{Z}_k + \dot{C}_{D,k})$. For the overall process, the relative cost difference shows that the cost per exergy unit of the product is 479% higher than the average cost per exergy unit of the fuel (delivered by coal and thermal energy). The exergoeconomic factor of 91% expresses the domination of the capital investment to the total cost associated with the overall process. The most important plant component from the economic and the exergoeconomic viewpoints is the electrolytic unit (plant component K). For both cost rates $\dot{Z}_k$ and $\dot{C}_{D,k}$, the highest values were calculated for the electrolyzer. More than 60% of the cost rate of the capital investment for the total process $\dot{Z}_{TOT}$ is associated with this single plant component. Regarding the results of the exergy analysis, the highest exergy destruction ratio was calculated, as expected, for the combustion reactor. However, the cost rate of exergy destruction in the electrolytic unit is by a factor 4 higher than the cost rate of exergy destruction in the combustion chamber. This is due to the fact that the unit cost of the fuel exergy for the electrolyzer is 20,67 €/GJ while it is only 2,33 €/GJ for the combustion chamber where the coal fuel is supplied to the process. Because of the high cost rates associated with the electrolyzer, the layout and the design parameters of this component should be chosen very carefully. An accurate modeling of this component is needed for the simulation and a careful estimation of the investment cost of this plant component is very important for the accuracy of the overall economic analysis. For the exergoeconomic optimization, the electrolyzer will always be the most important plant component.

In contrast to most of the plant components, plant component E (water separation and compression of the gases) has a relatively low exergoeconomic factor (19%). This is identifying the cost rate of exergy destruction to be the major cost source for this plant component. From the thermodynamic point of view, plant component E is very inefficient (ε_E = 20,7%, see Table 2) but it is needed for closing the thermodynamic cycle. About 10% of the electric power output of the gas turbine is required here for compressing the gases. Moreover, about 180 MW of low-temperature heat is released to the environment without any further usage. The increase of the unit costs between fuel and product for this component is 473% (Table 3). For the second compression group (plant component F), the exergoeconomic factor is relatively low, too. To optimize the efficiency and the cost effectiveness of the overall process, it should be further investigated whether (a) the entire compression process can be designed thermodynamically more efficiently, (b) replaced by other components, or (c) whether the separation process can be realized at a lower pressure, to avoid the high consumption of expensive electric energy in the compression process.

5 DISCUSSION

The advantages of the proposed process - being a true zero-emission process - can be summarized as follows:

- Hydrogen and electric energy are generated with an overall efficiency that is clearly higher than the efficiencies with which either hydrogen or electric energy could be generated from either carbonaceous solid fuels or high-temperature heat in any other known process.
- Thermodynamic losses of the combustion reaction are reduced when sulfuric acid is used as oxygen carrier.
- NO_X emissions are avoided since no air is needed in the combustion reaction.
- Solid fuels with high sulfur content can be used because of complete and easy SO_2 removal from the combustion gas within the process.
- The combustion product CO_2 is separated without any additional energy penalty and released as a compressed liquid for further transportation and storage.

Concerning further development of the energy conversion process discussed in this paper, experimental investigations are needed regarding the performance of the electrolyzer, the reaction kinetics in the combustion reactor, the gas cleaning at high temperature and pressure, and the gas turbine allowing expansion of gas mixtures including sulfur dioxide. A nuclear heat production plant was considered here as an example for the high-temperature heat source. For the operation of the process with high efficiency, the thermal energy must be available continuously on a high-temperature level. The results of the modeling of the process, of the economic, and the exergoeconomic analyses are very promising. However, an optimization should be the next step in the comprehensive evaluation of this novel process. The results of the exergoeconomic analysis will be used for an exergoeconomic optimization to reduce the product costs. A direct comparison with other existing or proposed energy conversion systems is difficult since here electric power and hydrogen are generated simultaneously based on two different energy inputs.

REFERENCES

[1] Tsatsaronis, G., 1999. "Strengths and Limitations of Exergy Analysis". In *Thermodynamic Optimization of Complex Energy Systems*, A. Bejan and E. Mamut, eds., Kluwer Academic Publishers, Netherlands, pp. 93–100.

[2] Knoche, K.F., Tsatsaronis, G., and Spindler, K., 1983. "HECAP: A New Process for the Production of Hydrogen and Electric Power" (in German), *Brennstoff-Wärme-Kraft*, **36**(11), pp. 530–534.

[3] Spindler, K., 1984. "Energy Conversion Processes with Sulfuric Acid used as Coal Oxidizer" (in German). PhD Thesis, Technical University of Aachen, Germany.

[4] Tsatsaronis, G., and Spindler, K., 1984. "Thermodynamic Analysis of a New Process for the Hydrogen and Electric Power Production by Using Carbonaceous Fuels and High-Temperature Process Heat", *International Journal of Hydrogen Energy*, **9**(7), pp. 595–601.

[5] Tsatsaronis, G., Kapanke, K., 2005. "Thermodynamic Analysis of a Novel Process with Integrated CO_2 Capture for the Production of Electric Power and Hydrogen". In Proceedings of IMECE 2005, ASME 2005 International Mechanical Engineering Congress and Exposition, November 5-11, 2005, Orlando, Florida USA, IMECE2005-79785

[6] Brecher, L.E., Spewock, S., Warde, C.J., 1977. "The Westinghouse Sulfur Cycle for the Thermochemical Decomposition of Water", *International Journal of Hydrogen Energy*, **2**(1), pp. 7–15.

[7] Schepers, H., 1983. "Thermodynamic Analysis of Sulfuric Acid Cycles for Water Splitting" (in German). PhD Thesis, Technical University of Aachen, Germany.

[8] Lahoda, E.J., Goossen, J.E., Matzie, R.A., Mazzoccoli, J.P., 2004. "Optimization of the Westinghouse Sulfur Process for Hydrogen Generation and the Interface with an HTGR". In Proceedings of ICONE12, 12th International Conference on Nuclear Engineering, April 2004, Arlington, Virginia USA, ICONE12-49031.

[9] Lahoda, E.J., 2005. "Research Opportunities for Improving the Westinghouse Sulfur Process (Hybrid Sulfur Process-HyS)". MIT Symposium on The Next Generation Nuclear Plant Goals and Challenges, February 23-24, 2005.

[10] Aspen Plus 11.1, 2001. See also: http://www.aspentech.com.

[11] Tsatsaronis, G. and Cziesla, F., 2002. *Encyclopedia of Physical Science and Technology*, 3rd edition, Vol. **16**. Academic Press, ch. Thermoeconomics, pp. 659–680.

[12] Diedrichsen, C., 1991. *Reference Environments for the Calculation of Chemical Exergy* (in German). VDI Verlag, Düsseldorf, Number 50 in Fortschr.-Ber. VDI, Reihe 19.

[13] Lazzaretto, A., Tsatsaronis, G., 1999. "On the Calculation of Efficiencies and Costs in Thermal Systems". In Proceedings of the ASME Advanced Energy Systems Division, AES-Vol. **39**, pp. 421–430.

[14] Bejan, A., Tsatsaronis, G., Moran, M., 1996. *Thermal Design and Optimization*, New York: John Wiley & Sons.

[15] Peters, M.S., Timmerhaus, K.D., 2003. *Plant Design and Economics for Chemical Engineers*. 5th edition, New York: McGraw-Hill.

ANALYSIS OF A NUCLEAR-POWERED AMMONIA PLANT

*William L. Kubic, Jr.** *and Andrei V. Belooussov*

Decisions Applications Division
Los Alamos National Laboratory
Los Alamos, New Mexico 87545

INTRODUCTION

Ammonia is an important industrial chemical and a potential energy carrier. Ammonia is 18% hydrogen by weight and it meets the 2015 freedom car energy density goal of 2.7 kWh/L. It can be used directly as a fuel in an internal combustion engine or an ammonia fuel cell, or it can be reformed into hydrogen for use in an internal combustion engine or hydrogen fuel cell. Ammonia has twice the energy density of liquid or compressed hydrogen, and experience shows that it can be handled, transported, and stored safely.

The convention process for producing ammonia is based on steam reforming of natural gas. However, domestic natural gas supplies are inadequate if ammonia is to be used as an alternative transportation fuel. Therefore, we are evaluating a nuclear-powered process as a carbon-neutral source of ammonia. Such a process is attractive because a nuclear reactor is continuous source of power and the potential exists for expanding the use of nuclear energy.

PROCESS FLOWSHEET

A conventional ammonia process consists of five phases: (1) manufacture of pure reactant gas, (2) compression, (3) catalytic reaction, (4) ammonia recovery, and (5) recirculation of reactants.[1] In a steam-reforming ammonia plant, reactant gas, which is a mixture of hydrogen and nitrogen, is produced by a series of reactions involving natural gas, steam, and air. A nuclear-powered ammonia plant consists of the same five phases. However, a very different process would be used to produce reactant gas. We envision a process in which reactant gas is produced by mixing pure hydrogen obtained from a water-decomposition process with nitrogen obtained from an air separation plant. The three major questions in developing a flowsheet for a nuclear-powered ammonia plant are as follows:

- Which process should be used for producing hydrogen?
- Which process should be used for producing nitrogen?
- Which type of nuclear power system should be used?

Hydrogen Production

Hydrogen production would be the major energy consumer in a nuclear-powered ammonia process. The efficiency of the hydrogen processes is a key parameter in determining the size of the nuclear reactor needed to power the process. Therefore, the efficiency of the hydrogen process has a major impact on both capital and operating costs.

---

* Author for corresponding (e-mail: wkubic@lanl.gov).

Four methods exist for hydrogen production: water electrolysis, steam electrolysis, thermochemical cycles, and hybrid cycles. Water electrolysis is existing technology for producing pure hydrogen. This technology could be powered by existing nuclear reactors, but the overall efficiencies would be low. Steam electrolysis is a promising new technology that could be more efficient than conventional water electrolysis. Efficiencies of 45 – 55 % have been reported by Idaho National Engineering and Energy Laboratory (INEEL) and the Japanese[2] for a processes powered by a high-temperature gas-cooled reactor (HTGR).

A thermochemical cycle is a series of thermally drive chemical reactions that split water into hydrogen and oxygen. Reactants in the process are regenerated and recycled. The iodine-sulfate cycle is perhaps the best known thermochemical cycle,[3] but many other cycles have been proposed.[4,5] Hybrid cycles are thermochemical cycles that include an electrochemical reaction step. The sulfate cycle is the most extensively studied hybrid cycle.[6] Theoretical efficiencies of 50 – 65% have been reported in the literature for thermochemical and hybrid cycles. However, theoretical estimates often neglect the energy consumed by the separation process. Integrated process studies indicate that more realistic estimates for the efficiencies of thermochemical and hybrid cycles are 40 – 45% (Ref. 3 and 6).

Our primary choice for hydrogen production is steam electrolysis. The available literature indicates that the practical efficiencies of steam electrolysis are comparable to the practical efficiencies of thermochemical processes. We also considered water electrolysis as a possible option. Although it is less efficient than steam electrolysis, it is proven technology.

Nitrogen Production

Commercial ammonia production requires large volumes of high-purity nitrogen. The reactant gas must contain <150 ppm of water and <10 ppm of oxygen and oxygen containing compounds, such as carbon dioxide. Three possible methods of producing large quantities of nitrogen exist: pressure-swing adsorption (PSA), cryogenic air separation, and burning hydrogen to remove oxygen.[7] Of these methods, PSA consumes the least amount of energy. The molecular sieve used for PSA removes carbon dioxide, but the nitrogen product typically contains 0.1 – 2 % oxygen. Commercially, high purity nitrogen can be obtained from a PSA plant by reacting the residual oxygen with hydrogen.

Combining nitrogen with hydrogen is an integral part of a nuclear-powered ammonia plant. High-purity reactant gas can be obtained by combining nitrogen from a PSA plant with hydrogen and passing the mixture through a catalytic de-oxo reactor. The residual water can be removed in a drier bed. Although this process consumes some hydrogen, PSA combined with a catalytic reaction of hydrogen and oxygen consumes significantly less energy overall than the other options. Therefore, we selected PSA as the preferred method of producing nitrogen.

Nuclear Power Source

Current commercial nuclear technology is limited to pressurized water reactors (PWRs) and boiling water reactors, which can provide heat at a maximum temperature of ~300°C. The US Department of Energy is considering several advanced reactor concepts: a very high temperature reactor, a gas-cooled fast reactor, a supercritical-water reactor, a sodium-cooled fast reactor, a lead-cooled fast reactor, and a molten-salt reactor.[8] The very high temperature reactor and a gas-cooled fast reactor are HTGRs and are capable of supplying heat at temperatures >850°C. Two HTGR prototypes are currently operating, which demonstrates that the concept is technologically feasible. The Japan Atomic Energy Research Institute operates the 30-MW High Temperature Engineering Test Reactor (HTTR),[9] which produces an outlet coolant temperature of 950°C, and the Institute of Nuclear Energy Technology, Tsinghua University, China operates a 10-MW pebble-bed reactor, which produces an outlet temperature of 700°C.

Steam electrolysis consumes large amounts of electricity so the efficiency of generating electricity was the primary consideration in selecting the nuclear power system because an HTGR with a Brayton cycle is our primary choice for the nuclear power system because an HTGR has the highest operating temperatures, which favors high cycle efficiencies. We also evaluated a PWR with a Rankine cycle as a possible option. Although it is less efficient than an HTGR, it is proven technology.

Baseline Process Design

We considered several alternative flowsheets in our evaluation. The baseline process design utilizes steam electrolysis, PSA, and an HTGR with a Brayton cycle. Figures 1 – 4 are simplified flowsheets for the baseline design.

Figure 1 shows the steam electrolysis process. The electrolytic cell produces pure oxygen and a hydrogen-steam mixture. We assume a single pass conversion of 50% (Ref. 2). To increase overall conversion and to reduce energy consumption, we incorporated a steam recycle loop. Heat from the hydrogen and oxygen product streams must be recovered to obtain the efficiencies reported in the literature.

Figure 2 shows the PSA process. We assume a typical operating pressure of 7.5 atm for the absorber beds. One difference between Fig. 2 and a commercial PSA is that we mix the wet hydrogen from the steam electrolysis process with the nitrogen to produce an impure reactant

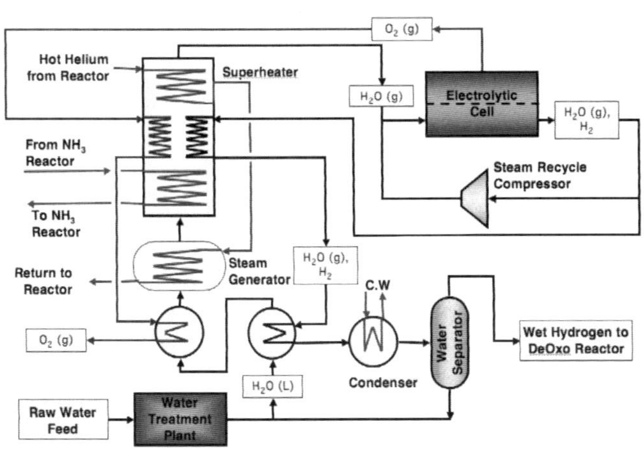

Fig. 1. Flowsheet for producing hydrogen with steam electrolysis.

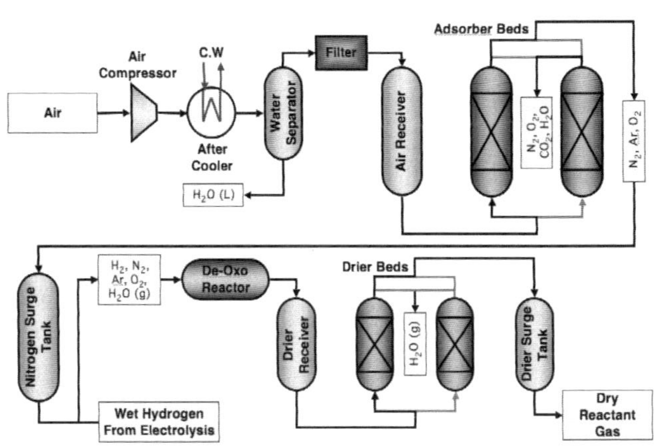

Fig. 2. Flowsheet for PSA.

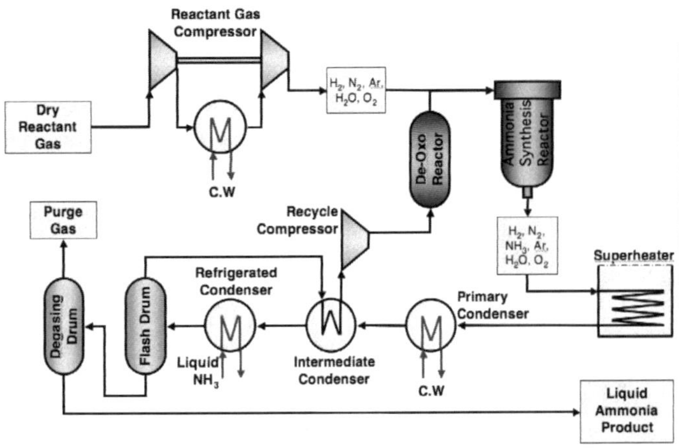

Fig. 3. Flowsheet for ammonia synthesis loop.

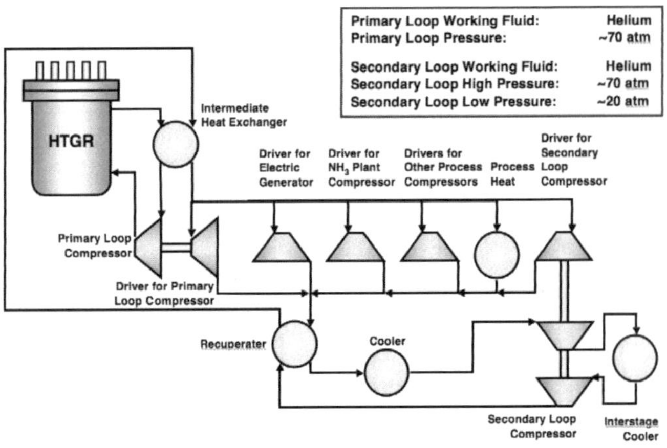

Fig. 4. Flowsheet for Brayton cycle.

gas, which is passed through re-oxo reactor to remove the residual oxygen. A commercial PSA plant would use a nearly stoichiometric quantity of hydrogen to remove residual oxygen.

Figure 3 is a simplified version of the flowsheet for the ammonia synthesis loop. The ammonia synthesis reaction takes place at typical temperature and pressure (i.e. 200 atm and 350 – 400°C), but the flowsheet differs from a conventional plant because the reactant gas contains less inerts than reactant gas produced by steam reforming natural gas. The baseline process uses an ammonia refrigeration system to condense the ammonia product at -25°C and 190 atm. At these process conditions, argon dissolves in the liquid ammonia and prevents excessive argon accumulation in recycle without a purge stream. Because the recycle stream is not purged, oxygen accumulation is prevented by passing the recycle gas through a catalytic reactor.

Figure 4 is a simplified diagram of the power distribution system. A two loop system is used to prevent radioactive contamination of the process equipment. Helium is the working fluid in both loops. The process uses a fully integrated power system to achieve maximum efficiency. Compressed, high-temperature helium powers the electrical generators, powers the process compressors, and supplies process heat.

PROCESS ANALYSIS

We performed material and energy balances for a 2000 tonne/day nuclear ammonia plant. We analyzed the baseline flowsheet illustrated in Fig. 1 – 4 as well as several alternatives. We also used pinch analysis to optimize the heat recovery network. Key assumptions used in the analysis are as follows:

- The electrolysis efficiency is 80% based on Gibbs free energy.
- Heat losses from the electrolyzers are negligible.
- The per pass conversion for steam electrolysis is 50%.
- The ammonia synthesis reactor is operated at 200 atm pressure.

- The exit stream from the ammonia synthesis reactor contains 20 mole % ammonia.
- The exit temperature from the HTGR is 900°C.

We defined the overall energy efficiency of the process as the net fuel value of the ammonia produced divided by the thermal energy generated in the nuclear reactor. Table 1 contains a summary of the efficiencies for the nuclear-powered processes considered in this study. Table 1 shows that the baseline design is the most efficient process with an overall efficiency of 40%. Generating electricity for hydrogen production consumes ~84% of the total energy, powering the process compressors consume ~10% of the energy and process heat consumes the remaining 6%. However, an efficiency of 40% can be achieved only with optimal heat recovery. A process with no heat recovery has an overall efficiency of only 34%, which is less than a process based on water electrolysis. The overall efficiency of a PWR-powered process is 21% which is about one-half that of the baseline design.

ECONOMIC ANALYSIS

We estimated capital costs and operating costs for the baseline HTGR-powered ammonia plant, the three alternative designs listed in Table 1, and two fossil-fuel based processes with and without carbon sequestration. All costs in this study are 2005 costs.

Assumptions for Estimating Costs

We separated the nuclear ammonia plant into two units for the economic analysis, the ammonia plant and the nuclear reactor. We used cost factors to estimate the capital costs for the ammonia plant from major equipment costs.[10] We obtained estimates for delivered equipment costs form a variety of sources and adjusted the costs to 2005 using the *Chemical Engineering* Plant Index. We estimated the capital costs of an HTGR by using a 0.7 power law to scale estimates for a modular helium reactor given in a General Atomics study.[11] We based the capital cost of a PWR on the Palo Verde Nuclear Power Station in Arizona.

TABLE 1. THE EFFICIENCY OF NUCLEAR-POWERED AMMONIA PLANTS

Reactor Type	Hydrogen Process	Energy Required (kwh$_t$/tonne)	Efficiency (MJ Fuel / MJ Thermal)
HTGR with Optimal Heat Recovery	Steam Electrolysis	12,600	0.40
HTGR	Water Electrolysis	14,200	0.35
HTGR with no Heat Recovery	Steam Electrolysis	14,700	0.34
PWR	Water Electrolysis	23,900	0.21

Operating costs for the ammonia unit can be divided into three major categories: energy costs, depreciation, and other costs. Other costs include labor, management, catalysts, and so forth. We assumed that these costs are equivalent to those of a conventional ammonia plant, and we obtained estimates of these costs from the literature.[12] We assumed that the operating costs for an HTGR are 0.55¢/kwh$_t$. This estimate is the average operating costs for commercial nuclear-power plants in the United States, and it is comparable to the estimated operating costs for a modular helium reactor.[11] We based operating cost of a PWR on the Palo Verde Nuclear Power Station.

Determining depreciation costs for a nuclear-powered ammonia plant is problematic because the depreciation periods for nuclear power plant and a chemical plant are very different. Typically, a nuclear power plant is designed for a 30 – 40 yr life, but the typical depreciation period for a chemical plant is 10 yrs. Because the nuclear reactor accounts for ~80% of the capital cost, we assumed a depreciation period of 30 yrs. To account for the shorter life of the ammonia unit, we assumed that it would undergo a major overhaul every 10 yrs to extend its useful life. We included these additional overhaul costs in the maintenance costs.

Cost Estimates for Nuclear-Powered Ammonia Production

The estimated equipment cost for the ammonia unit is $65.4 million. Figure 5 shows the equipment-cost breakdown. We estimated the total capital costs for the ammonia unit to be $270 million. The estimated capital cost for the HTGR is $1169 million, which accounts for 81% of the total capital cost of $1439 million.

Table 2 contains capital cost estimates for the four nuclear-powered options. These results indicate that steam electrolysis does not have a large capital-cost advantage over conventional water electrolysis. The difference in the total capital costs is within the uncertainty of the estimates. The results in Table 3 also show that the capital cost of an ammonia plant powered by a PWR is more than 50% greater than the capital cost of the baseline HTGR-powered process.

The estimated operating cost for the baseline design is $173/tonne. Figure 6 shows the breakdown of the operating costs. Table 3 contains the estimated operating costs for the other nuclear-powered options. Steam electrolysis does not have a large operating cost advantage over conventional electrolysis. The type of reactor used is a more important factor. Operating costs for an HTGR powered plant are significantly less than a PWR powered plant.

Comparisons with Other Options

We compared capital and operating costs of the nuclear-powered options with plants based on steam reforming of natural gas and partial oxidation of coal. The costs estimates for steam reforming were based on a natural gas price $8.00 per 1000 scf, and the cost estimates for partial oxidation were based on a coal price of $36 per short ton. Costs estimates for carbon sequestration were based on storing liquid carbon dioxide in a well 200 located km from the plant. Pipeline coasts were $0.02/tonne CO_2/km and disposal costs were $5.00/tonne CO_2 (Ref. 13).

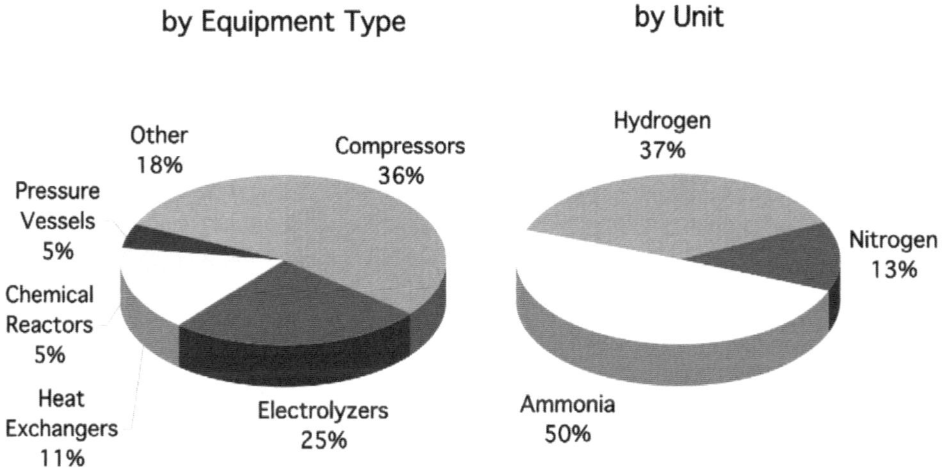

Fig. 5. Cost breakdown for the ammonia unit.

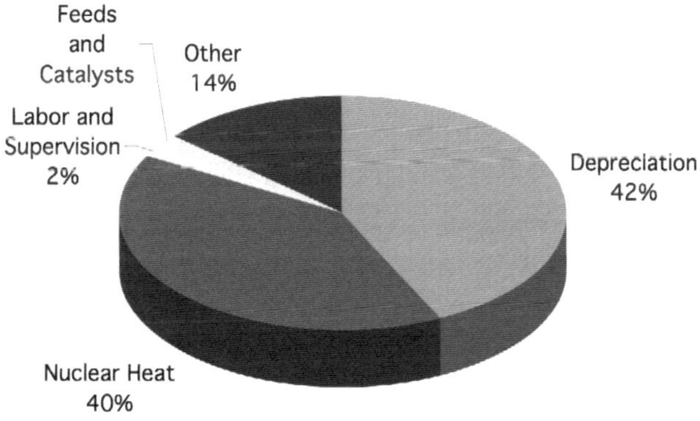

Fig. 6. Breakdown of operating costs for the baseline design.

TABLE 2. CAPITAL COST ESTIMATES FOR NUCLEAR-POWERED AMMONIA PRODUCTION

Reactor Type	Hydrogen Process	Ammonia Plant Cost (Million $)	Reactor Cost (Million $)	Total Cost (Million $)
HTGR with Optimal Heat Recovery	Steam Electrolysis	270	1169	1439
HTGR	Water Electrolysis	245	1274	1519
HTGR with no Heat Recovery	Steam Electrolysis	270	1303	1573
PWR	Water Electrolysis	245	2010	2255

TABLE 3. OPERATING COST ESTIMATES FOR NUCLEAR-POWERED AMMONIA PRODUCTION

Reactor Type	Hydrogen Process	Nuclear Heat ($/tonne)	Other Cost ($/tonne)	Total Cost ($/tonne)
HTGR with Optimal Heat Recovery	Steam Electrolysis	69	104	173
HTGR	Water Electrolysis	78	103	181
HTGR with no Heat Recovery	Steam Electrolysis	81	108	189
PWR	Water Electrolysis	131	132	263

Table 4 contains a summary of capital and operating costs for nuclear and fossil ammonia plants. Recent ammonia prices are included in the table for perspective. The nuclear-powered plants have the lowest operating costs but highest capital cost, while a natural-gas plant has the highest operating costs but lowest capital costs. Because of the inverse relationship between operating costs and capital costs, we compared options on the basis of return on investment (ROI).

Figure 7 shows the rate of return (ROR) as a function of ammonia price to the consumer for the options without carbon sequestration. Ammonia price is expressed in term of $ per gallon of gasoline equivalent (gge) and it includes $0.40/gge distribution costs and $0.45/gge taxes. The

TABLE 4. A COMPARISON OF COSTS FOR VARIOUS AMMONIA PRODUCTION OPTIONS

Type of Plant	Captial Cost (MM$)	Operating Cost ($/tonne)
Historic Average NH_3 Price	-	165
HTGR-Powered Plant (Bseline)	1439	173
Jan 2005 Tampa Barge Price	-	260
PWR-Powered Plant	2255	263
Coal Plant	868	265
Coal Plant with CO_2 Sequestration	996	311
Natural Gas Plant	362	343
Natural Gas Plant with CO_2 Sequestration	452	369
Jan 2006 Tampa Barge Price	-	374

target ROR for a chemical plant is typically 20 – 25%. Figure 7 shows that for RORs greater than 12% natural gas and coal have a price advantage over the best nuclear powered option. Figure 8 shows the ROR as a function of ammonia price to the consumer for the production options with carbon sequestration. For a 20% ROR, the HTGR-powered option has a slight price advantage over coal. However, natural gas still has the overall price advantage.

If the objective is producing ammonia for fertilizer, steam reforming natural gas is the best choice because it has a price advantage over other options. This price advantage holds even if the carbon dioxide is sequestered. However, if the objective is to produce a carbon-neutral fuel as a replacement for gasoline, the situation changes because the supply of natural gas becomes an issue. Domestic reserves of natural gas are limited, so other sources of ammonia would be needed. Our analysis indicates that a HTGR-powered plant may be the most economical means of producing ammonia as a carbon dioxide neutral fuel.

SUMMARY

An HTGR-powered ammonia plant has the lowest production costs of the options considered in this study. The estimated production cost is $173/tonne or $1.10/gge. However, the capital costs of a nuclear-powered plant are much greater than a plant utilizing natural gas or coal. The high capital cost is a significant economic obstacle to an economically viability of ammonia as a transportation fuel. To earn an adequate ROI, the retail price of ammonia produced by a HTGR-powered plant would need to be about $4.50/gge.[†]

Although the capital costs are high, a nuclear-powered plant may be the best means of ammonia production without carbon dioxide emissions. Producing ammonia from natural gas is

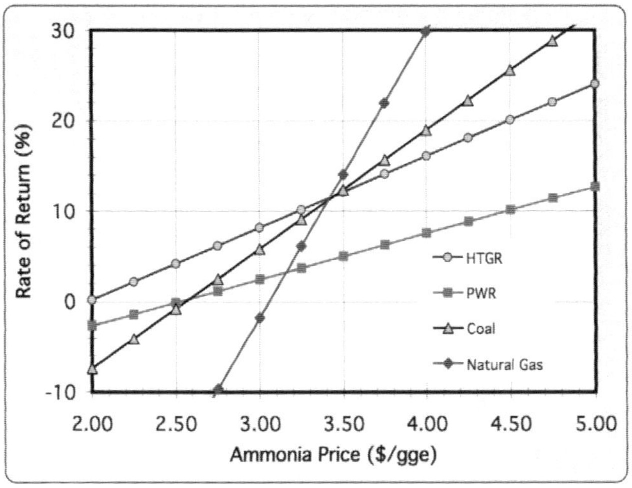

Fig. 7. Rate of return for nuclear and fossil ammonia plants without carbon sequestration.

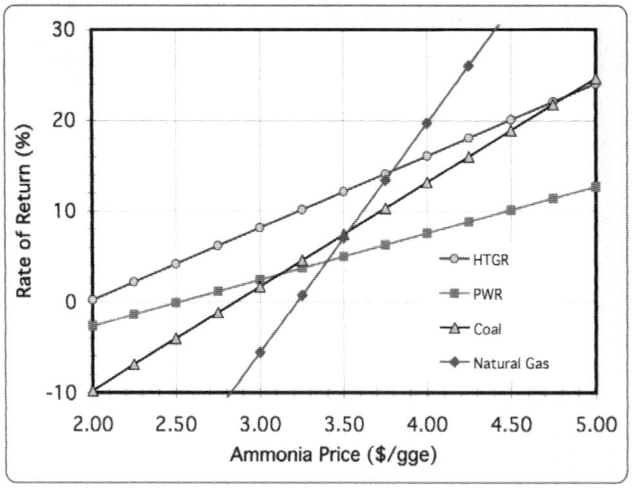

Fig. 8. Rate of return for nuclear and fossil ammonia plants with carbon sequestration.

[†] Incudes 20% ROI, $0.40/gge distribution costs, and $0.45/gge taxes.

not viable option because domestic reserves are inadequate for meeting US transportation needs. Domestic coal reserves are adequate, but a nuclear-powered ammonia plant has a slight economic advantage over a partial-oxidation plant with carbon sequestration.

REFERENCES

1. G. Austin, *Shreeve's Chemical Process Industries*, 5th Ed., McGraw-Hill Book Company, New York (1984).

2. R. Hino, K. Haga, H. Aita, and K. Sekita, "R&D on Hydrogen Production by High-Temperature Electrolysis of Steam," *Nuclear Engineering and Design*, 233: 363 – 375 (2004).

3. M. Nomura, S. Nakao, S. Kasahara, K. Ikenoya, S. Kubo, and K. Onuki, "Development of an Electrochemical Cell for Efficient Hydrogen Production through the IS Process," *AIChE Journal*, 50: 1991 – 1998 (2004).

4. K-S Sim, Y-M Son, and J-W Kim, "Some Thermochemical Cycles Composed of Copper Compounds with Three-Step Reactions," *International Journal of Hydrogen Production*, 18: 287 – 290 (1993).

5. C. Forsberg, "Hydrogen, Nuclear Energy, and the Advanced High-Temperature Reactor," *International Journal of Hydrogen Energy*, 28: 1073 – 1081 (2003).

6. E. Bilgen and C. Bigen, " A Hybrid Thermochemical Hydrogen Producing Process Based on the Cristina-Mark Cycles," *International Journal of Hydrogen Energy*, 11: 241 – 256 (1986).

7. T. Hardenburger and E. Ennis, "Nitrogen," in *Kirk-Othmer Encylopedia of Chemical Technology*, 5th Ed., John Wiley & Sons, Inc., New York (2005).

8. "Six Generation IV Technology Concepts Selected for R&D," Idaho National Laboratory web site http://energy.inel.gov/gen-iv/default.shtml.

9. H. Ohashi et al., "Performance Test Results of Mock-up Test Facility of HTTR Hydrogen Production System," Journal of Nuclear Science and Technology, 41: 385–392 (2004).

10. F. Holland and J. Wilkinson, "Process Economics," in *Perry's Chemical Engineers' Handbook*, 7th Ed., R. Perry and D. Green, Eds, McGraw-Hill Book Company, New York (1997).

11. D. Albertein, "Plutonium Consumption Modular Helium Reactor (PC-MHR) Plutonium Disposition Study Summary Report," General Atomics report GA-A22367 (January 1996).

12. T. Eggman, "Ammonia," in *Kirk-Othmer Encylopedia of Chemical Technology*, 5th Ed., John Wiley & Sons, Inc., New York (2005).

13. M. Gambini and M. Vellini, "CO2 Emission Abatement from Fossil Fuel Power Plants by Exhaust Gas Treatment," *Journal of Engineering for Gas Tubines and Power*, 125: 365 – 373 (2003).

Aspen Plus^{TM} Process Design for the Liquid Hydrogen Production by Steam Reforming of Used Automotive Lubricating Oil

*Mohamed M. Elbaccouch, Karthikeyan K.Ramasamy, Jong Baik, and Ali T-Raissi**

University of Central Florida
Florida Solar Energy Center
1679 Clearlake Road
Cocoa, FL 32922

Tel: (321) 638-1446
Fax: (321) 504-3438
Email: ali@fsec.ucf.edu

** Corresponding author.*

Abstract

The objective of this study is to develop a process for the production of pure liquid hydrogen from used automotive lubricating oil using Aspen Plus^{TM} Chemical Process Simulator (CPS). The reformer is simulated as a Gibbs reactor to fractionate the large oil molecules. A methane reformer converts methane into carbon monoxide and hydrogen. A water-gas shift unit, simulated as an equilibrium reactor, is used to enhance the production of hydrogen. A flash separator is used to knock out excess water from the product stream. Aspen Adsim^{TM} is incorporated to simulate the adsorption process of the pressure swing adsorption (PSA). The off-gas is burned with air in a combustor, and the generated energy is used to heat the processing units. A hydrogen liquefaction unit, simulated as a Claude densifier cycle, is used to convert gaseous hydrogen to liquid hydrogen. The performance of plant is investigated as a function of temperature and steam-to-oil ratio.

Introduction

Billions of gallons of used lubricating oil are available through out the year in the United States at approximately 10 cents per gallon of delivered cost [1]. Used lubricating oil is a high volume resource with a high energy density that can be converted into a valuable energy source. Elemental analysis of virgin synthetic and used lubricating oil contains 13.4 and 13.7 wt% hydrogen respectively. Calculations have shown that one gallon of used automotive lubricating oil produces approximately 1.38 kg of hydrogen via steam reformation process. Used lubricating oil has an approximate density of 0.88 $g.ml^{-1}$ at 20 ^{o}C and consists typically of 73-80 wt% aliphatic hydrocarbons, 11-15% monoaromatic hydrocarbon, 2-5% diaromatic hydrocarbons, and 4-8% polyaromatic hydrocarbons [1].

The objective of this work is to simulate an economically viable process for the production of liquid hydrogen by steam reforming used automotive lubricating oil. The simulation is carried out using Aspen Plus^{TM} Chemical Process Simulator (CPS). This project is part of an effort aimed at simulating chemical plants for the production of hydrogen via a variety of technologies and feed stocks [2-5].

Aspen Plus^{TM} Process Model

Figure 1 is a schematic diagram of the Aspen Plus^{TM} flowsheet for the steam gasification of used oil to produce pure liquid hydrogen. The process operates at atmospheric pressure and consists mainly of a Gibbs reactor for the thermal cracking of used oil with steam followed by a methane reformer to produce carbon monoxide and hydrogen ($CH_4 + H_2O \leftrightarrow CO + 3H_2$). Then, a water-gas shift reactor enhances further the hydrogen productivity ($CO + H_2O \leftrightarrow CO_2 + H_2$) [6]. The product stream is then introduced to a flash separator to remove excess water. A pressure swing adsorption (PSA) unit recovers pure hydrogen from the product stream. A combustor converts the off-gas into energy which used to heat processing units. The pure gaseous hydrogen product stream is introduced to a Claude Cycle where it is converted to liquid hydrogen. All of the unit operations in the plant run adiabatically with the exception to the Gibbs reactor whose temperature was pre-set. The off-gas stream is used to sustain the adiabatic requirement of the plant. The gaseous hydrogen production section of the model is simulated using the simple Peng Robinson cubic equation of state with the Boston-Mathias alpha function (PR-BM).

The Gibbs reactor minimizes the total Gibbs free energy of the feed streams to the reactor assuming that all of the components defined in the simulation in the component specification sheet are viable products [2]. The Gibbs reactor reforms the complex oil mixture by fractionating it down into simple hydrocarbon molecules, such as H_2, H_2O, CO, and CO_2, and CH_4. The adsorption capability of the PSA unit is simulated with the aid of Aspen-Adsem^{TM}. 15 mol% of the pure hydrogen produced from the PSA is recycled in order to meet the 70 mol% hydrogen requirements in the feed stream to the PSA [7,8].

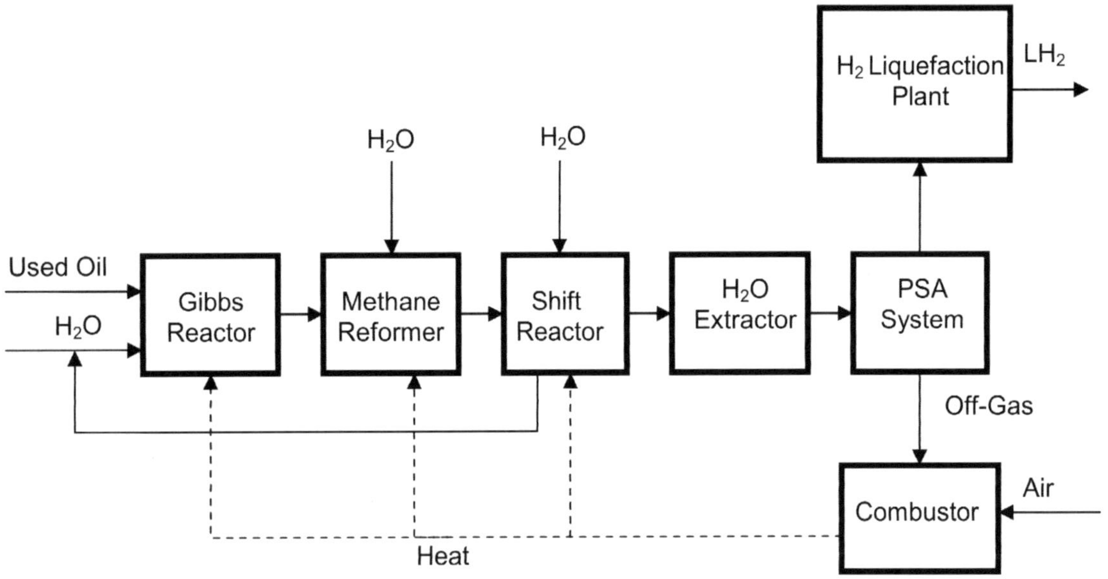

Figure 1. Aspen Plus^{TM} flowsheet for liquid hydrogen production by steam reformation of used automotive lubricating oil.

The gaseous hydrogen stream produced from the PSA unit is directed into a hydrogen liquefaction unit shown schematically in Figure 2. The temperature and pressure specifications of the heat exchangers are given in the figure. The hydrogen liquefaction unit uses liquid nitrogen pre-cooled Claude Cycle with an isentropic expander and an isenthalpic expansion valve. A multi-stage hydrogen compressor with intercoolers produces 80 atm of high pressure gas while the temperature remains at room temperature, and the liquid nitrogen bath cools the compressed gas down near 77 K [9]. Some of the compressed gas is routed to the expander, and then it gets colder by experiencing isentropic expansion process. The expanded gas helps to cool the counter flow high pressure gas in the heat exchanger. After the high pressure gas gets colder through the coldest heat exchanger, the cold gas experiences isenthalpic expansion process over the Joule-Thompson expansion valve. The expansion valve generates 20.3K of liquid-vapor mixture of hydrogen at 1 atm. The liquefied hydrogen is separated to storage tanks and only the saturated vapor goes back to the liquefaction cycle. The same amount of makeup gas as the liquefied hydrogen is supplied to the liquefaction cycle to complete the cycle. The hydrogen liquefaction unit is modeled using the Benedict-Webb-Rubin-Lee Starling (BNVR-LS) equation of state [6].

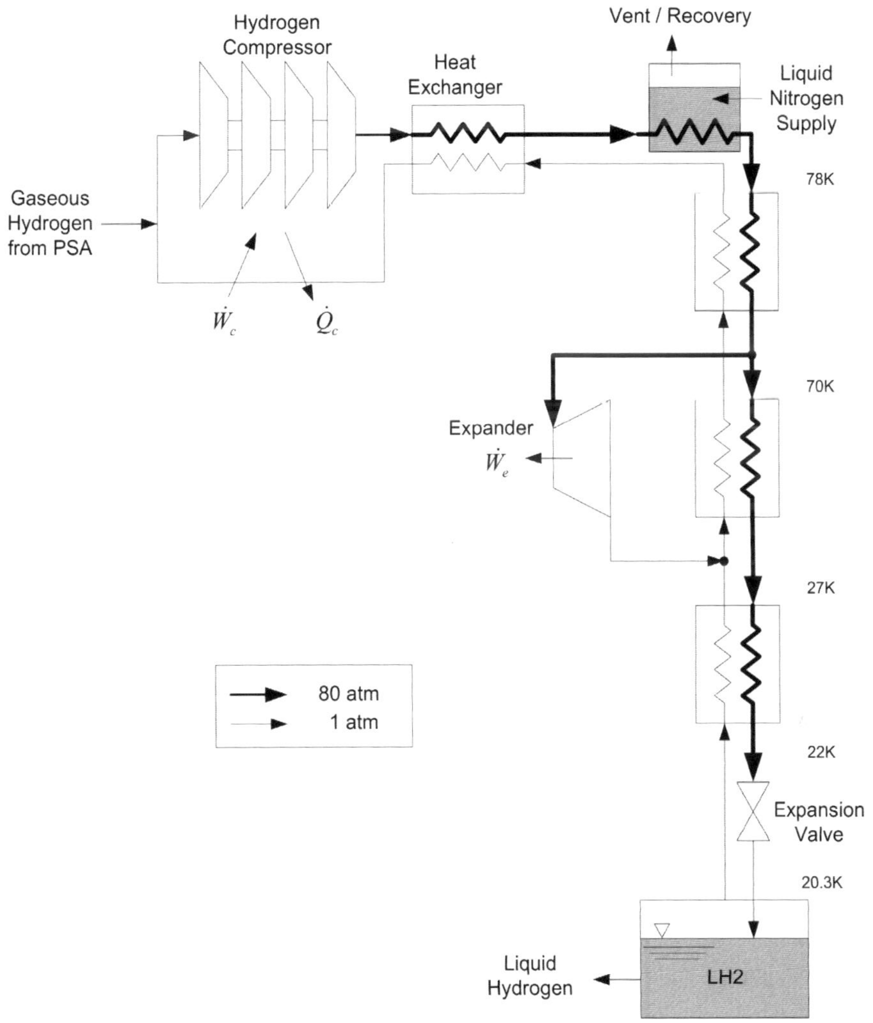

Figure 2. Schematic diagram of the Aspen Plus™ Claude Cycle hydrogen densifier.

Results and Discussions

Table 1 shows the optimized unit operation parameters for the Aspen flowsheet. The temperature and steam inlet to the Gibbs reactor are held constant at 650°C and 11.5 kmol.hr^{-1} respectively while varying the used oil flow rate. Most of the hydrogen is produced in the Gibbs reactor. In addition, the water-gas shift reactor enhances the hydrogen production. The hydrogen in the off-gas feed stream was burned with air to generate heat for the adiabatic operation of the model. Table 2 depicts hydrogen production as a function of the Gibbs reactor temperature while holding the used oil input constant at 1 kmol.hr^{-1}. Also, it shows that there is a slight increase in hydrogen flow rate as the temperature of the Gibbs reactor increases from 650 to 950 °C. Reducing the steam input to the reformer enhances hydrogen production.

Table 1. Optimized unit operation parameters for gaseous hydrogen production for a Gibbs reactor at 650 °C and 11.5 $kmol.hr^{-1}$ steam.

Gibbs react temp	°C	650	650	650	650
Oil Gibbs react feed stream	$kmol.hr^{-1}$	1	4.4	7.8	11.3
H_2O Gibbs react feed stream	$kmol.hr^{-1}$	11.5	11.5	11.5	11.5
H_2 Gibbs react prod stream	$kmol.hr^{-1}$	11.4	24.9	39.1	53.4
H_2 Shift react prod stream	$kmol.hr^{-1}$	3.8	4.3	4.7	5.0
H_2 PSA react prod stream	$kmol.hr^{-1}$	12.5	24.8	37.2	49.7
Total off-gas inlet stream	$kmol.hr^{-1}$	10.2	14.5	19.6	24.0
H_2O shift react feed stream	$kmol.hr^{-1}$	20.0	7.0	7.0	7.0
H_2 off-gas feed stream	$kmol.hr^{-1}$	2.7	4.4	6.6	8.8
H_2 produced from combustor	$kmol.hr^{-1}$	0.003	0.004	0.007	0.009
Shift React Prod temp	°C	213	198	198	198
Combustor React temp	°C	1368	1660	1640	1909
Shift React Press	atm	20	15	15	15

Table 2. Optimized unit operation parameters as a function of Gibbs reactor temperature with one $kmol.hr^{-1}$ used oil input.

Gibbs React Temp	Unit	650°C	700°C	750°C	850°C	950°C
Oil Gibbs react feed stream	$kmol.hr^{-1}$	1	1	1	1	1
H_2O Gibbs react feed stream	$kmol.hr^{-1}$	11.5	10.0	8.5	7.5	7.5
H_2 Gibbs react prod stream	$kmol.hr^{-1}$	11.4	12.1	12.2	12.5	12.9
H_2 shift react prod stream	$kmol.hr^{-1}$	3.8	5.1	6.1	7.0	7.2
H_2 PSA react prod stream	$kmol.hr^{-1}$	12.5	14.7	15.5	16.5	17.1
Total off-gas feed stream	$kmol.hr^{-1}$	10.2	10.1	10.3	10.5	10.6
H_2O shift react feed stream	$kmol.hr^{-1}$	20.0	25.0	27.4	30.0	31.0
H_2 off-gas feed stream	$kmol.hr^{-1}$	2.7	2.6	2.7	3.0	3.0
H_2 produced from combustor	$kmol.hr^{-1}$	0.003	0.003	0.003	0.003	0.003
Shift react prod stream	°C	212	212	215	215	215
Combustor react temp	°C	1368	1068	949	806	730
Compressor press	atm	20	20	21	21	21

It is relevant to perform a sensitivity analysis on the Gibbs reactor (reformer) since most of the hydrogen is produced in the reformer. Also, the reformer acquired positive heat input, while the remaining processing units are auto-thermal meaning that their heat input is generated from combusting the off-gas stream. For an efficient and optimized process design, the model requires that the plant does not generate soot (carbon). At a given reformer temperature and pressure, there is a definite steam-to-oil molar ratio (XW) above which no carbon forms in the Gibbs reactor. The theoretical desired state for the reformer operation is the one where the concentration of carbon, carbon dioxide, and water species are equal to zero. Figure 3 represents the equilibrium concentration of the major products from the reformer (1 atm, 650 °C, and 1 $kmol.hr^{-1}$ used oil) as a function of the normalized molar concentration of the input steam to the reformer. Note that above 0.9 the concentration of carbon is zero, and the concentration of carbon dioxide, carbon monoxide, and water diminish significantly.

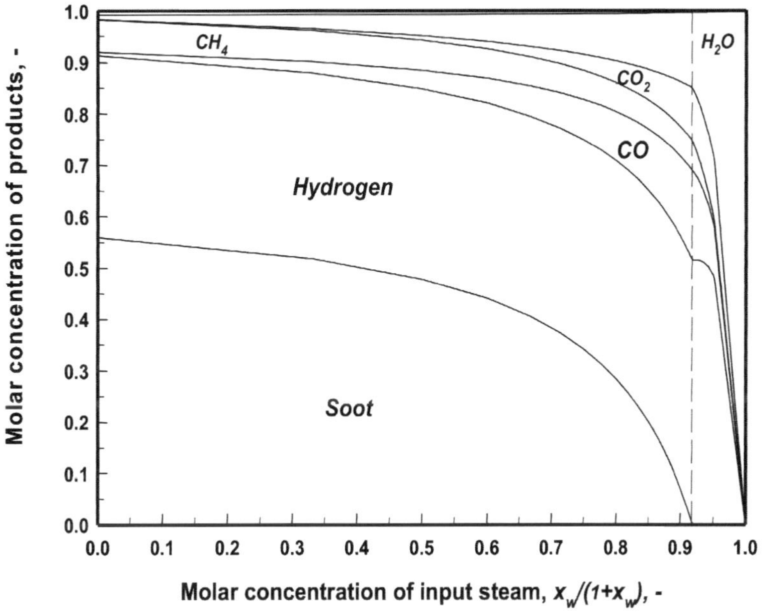

Figure 3. Equilibrium concentration of Gibbs reformer products as a function of the normalized input steam to the reformer at 1 atm, 650°C, and 1 kmole.hr^{-1} used oil.

Conclusions

This study utilizes Aspen Plus™ to design a plant for the production of liquid hydrogen by the steam reformation of used lube oil. The used oil inlet to the reformer is varied at a constant reformer temperature and steam input of 650 °C and 11.5 kmol.hr^{-1} respectively. Varying the temperature of the reformer (650-900 °C) while keeping the used oil input constant at one kmol.hr^{-1} does not have a significant impact on hydrogen production. Most of the hydrogen is produced within the reformer. Also, the methane reformer as well as the water-gas shift reactor enhance the production of hydrogen. The plant generates its energy auto-thermally via burning the off-gas stream with air. The reformer is the only unit operation that requires an endothermic energy.

Sensitivity analysis of the reformer at 650 °C, 1 atm, and 1 kmole.hr^{-1} used oil shows that operating the plant above 0.9 ratio of steam-to-oil is the optimum ratio because the reformer does not produce carbon, and had the minimum output of CO, CO_2, and H_2O. Claude cycle and Aspen Adsim™ data will be presented to examine the adsorption-desorption capability of the PSA unit, and energy consumption of the hydrogen liquefaction cycle.

Acknowledgements

The authors acknowledge the support provided by the National Aeronautics and Space Administration through Glenn Research Center under contract No. NAG32751.

References

[1] Used Mineral-Based Crankcase Oil, Chemical and Physical Information. http://www.atsdr.cdc.gov/toxprofiles/tp102-c3.pdf.

[2] Elbaccouch MM, and T-Raissi A. Aspen Plus Process Model for the Production of Gaseous Hydrogen via Bagasse Gasification. In the Proceedings of the *15th World Hydrogen Energy Conference*, Yokohama, Japan, 2004.

[3] Muradov NZ, T-Raissi A, and Robertson T. Hydrogen Production via Catalytic Reformation of Low-Quality Methane Containing Feedstocks. Presentation at the

Hydrogen Power Theoretical and Engineering Solutions International Symposium (HYPOTHESIS), Porte Conte, Italy, 2003.
[4] Kinoshita CM, and Turn SQ. Production of Hydrogen from Bio-Oil Using CaO as a CO_2 Sorbent. *Int J Hydrogen Energy 28(10) (*2003) 1965-2003.
[5] Bain A. NASA Kennedy Space Center Base Center Hydrogen Operations. Florida Solar energy Cente, University of Central Florida, FSEC- CR-1359-02, October 2002.
[6] Skolnik E, and Putsche V. Hydrogen Implementation Agreement, Task 11: Integrated Systems, Final Report of Subtask B: Analysis Tools. International Energy Agency, 1999.
[7] Kratz WC, Rarig DL, and Pietrantonio JM. Hydrogen and Carbon Dioxide Coproduction from SMR OFF Gas by Pressure Swing Adsorption. *AICHE Symposium Series 84(264) (1*988) 36-43.
[8] Ruthven DM. Principles of Adsorption and Adsorption Processes. John Wiley & Sons, New York, 1984.
[9] Klaus D. Timmerhaus KD, and Flynn TM. *Cryogenic Process Engineering,* Plenum Press, New York, 1989.

Solid-state Hydrogen Storage System Development and Engineering Analysis

T. Motyka and D. L. Anton
Energy Security Department
Savannah River National Laboratory
Washington Savannah River Co.
Aiken, SC, 29808
E-mail: ted.motyka@srnl.doe.gov

Abstract:

The development of a viable hydrogen storage system is one of the key challenges that must be met prior to the establishment of a true hydrogen economy. Current hydrogen storage options, such as compressed gas and liquid hydrogen, fall short of meeting vehicle manufacturers' goals for safe and efficient energy storage. The most viable long-term alternative to these options is solid-state storage, which has been proven both safe and efficient. The Savannah River National Laboratory (SRNL), with over 50 years of hydrogen storage expertise and over 25 years of expertise in solid-state storage, is working with the Department of Energy's Hydrogen Storage Center of Excellence in Metal Hydrides to help solve this key challenge.

A considerable amount of the Center's current efforts involve research to develop new high capacity hydrogen storage materials. However the overall objective of the Center is to deliver a 1 kg of hydrogen prototype vessel that meets the 2010 U.S. DOE FreedomCAR targets, which includes 6wt% hydrogen in the system. To meet this challenging target, will not only require the development of a new light weight hydrogen storage material but also, will require similar breakthroughs in vessel and system design, engineering and fabrication technology. SRNL is leading this systems engineering and analysis activity for the DOE Metal Hydride Center of Excellence. The SRNL team is comprised of distinguished scientists and engineers from national laboratories, leading universities, and major corporate research centers that have extensive experience in solid-state hydrogen storage systems as well as supporting expertise in heat transfer, systems modeling and component and systems development and manufacturing.

To meet the FreedomCAR storage goals, engineering and system development tasks will need to be worked in parallel with material development. This approach will ensure that material development and engineering efforts are coordinated and focused on creating a suitable storage material and a compatible storage container. This approach considers the many required tradeoffs. For example, selection of a hydrogen storage material with a high equilibrium pressure could adversely affect the vessel's pressure design and, in turn, its weight, and perhaps the overall hydrogen charging time. Lower pressure materials would permit the use of lightweight storage containers and even possible conformable tank shapes.

The Center's efforts will build on the experience and ongoing work of its lead and partners. The Center will also share its knowledge and expertise to solve emerging safety and engineering issues. The major engineering and system development tasks being carried out by the center include:

- Heat transfer modeling and verification
- Engineering materials property measurements i.e. kinetics, heat capacity, packing density, cycling performance etc.
- Material production scale-up.
- Safety issues including risk identification and mitigation.
- Subsystem fabrication and testing
- Prototype system fabrication and testing
- System modeling and integration.
- Manufacturability and cost estimates.

The accompanying presentation will describe the current hydrogen storage systems engineering challenges in more detail and the approach that the SRNL-led team will follow to help solve these challenges.

256b Vehicular Hydrogen Storage: Goals, Challenges, and Progress

Grace Ordaz[1], **John Petrovic**[2], Carole Read[1], George Thomas[3], and Sunita Satyapal[1]. (1) Office of Hydrogen, Fuel Cells and Infrastructure Technologies, U.S. Department of Energy, 1000 Independence Ave., SW, Washington, DC 20585, (2) Los Alamos National Laboratory, 8954 Puerto Del Rio Drive, #402, Cape Canaveral, FL 32920, (3) Sandia National Laboratories, 1000 Independence Ave., SW, Washington, DC 20585
Hydrogen is an energy carrier with the potential to replace gasoline for vehicular applications. However, hydrogen-powered vehicles require a driving range of greater than 300 miles in order to meet customer needs and effectively compete with other technologies. For the overall vehicular fleet, this dictates that substantial quantities of hydrogen be stored on-board within stringent weight, volume, and system cost constraints, which constitutes a major scientific and technological challenge. To meet this challenge, the DOE's National Hydrogen Storage Project has been formed which focuses on materials-based technologies. Centers of Excellence in metal hydrides, chemical hydrides, and carbon-based materials have been established, as well as independent university and industry projects in the areas of new concepts/materials, hydrogen storage testing, and storage system analysis. Recent technical progress in each of these areas will be presented as well as collaborative hydrogen storage activities under the International Partnership for the Hydrogen Economy (IPHE).

256c Bench Scale Saes St 909 Tests for Methane and Carbon Dioxide Removal from Helium, Hydrogen, and Nitrogen Gas Streams

James E. Klein, Savannah River National Laboratory, Aiken, SC 29808

SAES St 909, a Zr-Mn-Fe alloy, was developed for decomposing, or "cracking", tritiated water in fusion reactor effluents. St 909 has also been shown to crack methane in helium, but its ability to purify hydrogen streams was uncertain, especially with nitrogen impurities. To better characterize St 909 impurity removal performance, bench scale (six gram) screening tests were performed using ten whole St 909 pellets. Cold (non-radioactive) tests were conducted to determine the ability of St 909 to crack methane and carbon dioxide. The tests used up to 5 vol% carbon dioxide plus methane concentrations to examine the impact of methane decomposition at 700°C and 800°C at 101mPa (760 torr). Carrier gases were helium, helium with 2 vol% ammonia, hydrogen, and nitrogen at a total flow of 10 sccm. Methane cracking was most efficient in helium and less efficient in hydrogen, due to the reduced driving force for the methane decomposition reaction, and in nitrogen, due to nitrogen absorption by the St909. A reduced methane cracking rate in the presence of ammonia was attributed to nitrogen absorption and not the hydrogen released from ammonia cracking. Carbon dioxide decomposition was most efficient in hydrogen, followed by helium, then nitrogen. High carbon dioxide decomposition rates in hydrogen were due to not only carbon and oxygen gettering by the St909, but also by methane, then later, water formation reactions. Carbon dioxide decomposition in nitrogen was reduced by nitrogen absorption by the St909 interfering with the decomposition and diffusion processes. Methane decomposition rates were reduced when carbon dioxide was present. In a helium and a nitrogen carrier, carbon dioxide can effectively stop methane cracking. In hydrogen, carbon dioxide can be reformed into methane to increase the exit methane concentration above its feed concentration. The tests and proposed mechanisms showed carbon dioxide decomposing to carbon monoxide. The carbon monoxide produced was still preferentially decomposed over methane thus suppressing the over-all methane cracking rate.

Novel light weight complex hydrides for hydrogen storage

Sesha Srinivasan, *Elias Stefanakos*
Clean Energy Research Center, College of Engineering, University of South Florida, Tampa, FL 33620

Abstract

The development of light weight hydrogen storage systems with high volumetric and gravimetric hydrogen densities is indeed essential for realizing the pollution free vehicular transportation [1]. Though, conventional metal hydrides possess the highest volumetric densities, their gravimetric capacities are rather low (< 2wt.%). On the other hand, complex chemical hydrides bearing light weight elements such as Li, Na, Mg, B, Al etc. show promising hydrogen storage behavior at moderate temperatures. Some of the complex hydrides are having high theoretical capacity of 10.0-18.0 wt.% (e.g. $NaBH_4$, $LiBH_4$ etc). However, the reversibility of these hydrides require the following individual or combined effects, (i) catalyst doping, (ii) lattice substitution and (iii) high temperature and pressure operation
.
The breakthrough discovery of Ti- catalyzed $NaAlH_4$ with enhancement in the dehydrogenation and rehydrogenation behavior at moderate temperatures revolutionized the scientific community to re-visit the complex hydrides for hydrogen storage [2]. The available hydrogen storage capacity in Ti- doped $NaAlH_4$ is 5.6 wt.%, which is well below the DOE technical targets [3]. In the present investigation, we have developed the novel light weight complex hydrides and demonstrated their hydrogenation characteristics. A new class of hydrogen storage system $Zn(BH_4)_2$ with theoretical hydrogen capacity of 8.4 wt.% has been synthesized by an inexpensive mechanochemical process. The enthalpy of decomposition and the gravimetric capacity of these hydrides are measured using high pressure DSC and TGA techniques. The structure-property correlations of the as-prepared material have been analyzed via various metrological tools. We have also carried out the methods to destabilize the borohydride systems (e.g. $LiBH_4$) by incorporating MgH_2 for cyclic reversibility in presence of $ZnCl_2$ catalyst.

1. Introduction

Light weight, low cost, high reversible hydrogen storage systems are indeed essential for realizing the low temperature PEM fuel cells powered vehicles [4, 5]. The breakthrough discovery of Ti- catalyzed $NaAlH_4$ [6, 7] for the reversible onboard hydrogen storage seems may not be the ideal system to attain the DOE 2010 and FreedomCAR technical targets [8, 9]. This is because for $NaAlH_4$, the usable hydrogen storage capacity achievable is 5.6 wt.%, which is well below the set DOE goals. On the other hand, the borohydride complexes such as $NaBH_4$ and $LiBH_4$ possesses high hydrogen storage capacity of 13.0 wt.% and 19.6 wt.% respectively [10]. However, the release of hydrogen from $NaBH_4$ is possible only by hydrolysis (reaction with H_2O) and this process is irreversible [11]. For the case of $LiBH_4$, the catalytic addition of SiO_2, significantly enhances its thermal desorption at 200° C [12]. The incorporation of various metal oxides such as TiO_2, V_2O_5 etc. have been shown to improve the reversible hydrogen sorption characteristics of $LiBH_4$ [13]. In any case, the thermal decomposition and/or recombination of hydrogen either from $NaBH_4$ or $LiBH_4$ accompanies with the thermodynamic barrier and structural stability due to strong B-H interactions [14-17]. A new transition metal assisted complex borohydrides, $Zn(BH_4)_2$ is recently reported with the high theoretical hydrogen storage capacity of 8.4 wt.% at the thermal decomposition temperature of 85° C [17]. The preparation of $Zn(BH_4)_2$ via wet chemical synthesis route [18] or mechanical milling [19, 20] is being investigated according to the stoicheometric reaction,

$$2NaBH_4 + ZnCl_2 \longrightarrow Zn(BH_4)_2 + 2NaCl \qquad (1)$$

$Zn(BH_4)_2$ thermally decomposes into the constituent elements and releases hydrogen,

$$Zn(BH_4)_2 \longrightarrow Zn + 2B + 4H_2 \qquad (2)$$

The present paper aims to systematically study the synthesis and characterization of new Zn- based complex borohydrides, $Zn(BH_4)_2$ for the efficient hydrogen storage. The active hydrogen content of 8.4 wt.% have been obtained at 100° C for the hydride milled for only 30 minutes; this capacity however decreases with increasing milling duration. For the case of complex mixture, ($2LiBH_4+ZnCl_2$), the gravimetric capacity of 10.8 wt.% is achieved at temperature below 120° C. The thermal decomposition characteristics of $Zn(BH_4)_2$ exhibits an endothermic reaction process, with the ΔH value corresponds to -1 to -5 kJ/mol H_2 for the $NaBH_4$ precursor and -28 kJ/mol H_2 for the $LiBH_4$ precursor. The destabilization of the $Zn(BH_4)_2$ structure for the reversibility reactions have also been carried out by doping the complex hydride with 0.1 mol% MgH_2.

2. Experimental Details

2.1. Materials and method

Starting materials such as $NaBH_4$ (98%), $LiBH_4$ (99.99%), $ZnCl_2$ (99%) and TiF_3 (99%) are obtained from Sigma Aldrich and used without further purification. All chemical reactions and operations are performed under nitrogen filled glove box. $NaBH_4$ ($LiBH_4$) and $ZnCl_2$ with 2:1 mole ratio has been mixed in the stainless steel bowl (80 ml) and sealed the lid with viton O-ring under nitrogen filled glove box. The bowl was then evacuated for at least 1h to remove the residual oxygen and moisture down to ppm level. A specially designed lid (www.globalgilson.com) with an inlet and outlet valves are used for this purpose. The mechanochemical process employing high energy milling has been carried out by Fritsch pulversette planetary mono mill, P6 under argon atmosphere. The milling parameters such as ball to powder weight ratio and milling speed are optimized to 20:1 and 300 rpm respectively. Milling duration has been varied as 30 min., 1h, 2h, 4h and 8h. Thus mechanochemically processed complex hydrides are immediately transferred to the glove box for further characterization measurements. In a similar way, TiF_3 and $NaBH_4$ ($LiBH_4$) with 3.1 mole ratio have been prepared for comparing the performance.

2.2. PXD, FTIR, Micro-Raman Characterization

The powder x-ray diffraction of the mechanically milled complex borohydride has been carried out by the Philips X'pert diffractometer with CuKα radiation of λ= 5.4060 Å. The incident and diffraction slit width used for the measurements are 1° and 2° respectively. The PXD phase identification and particle size calculation has been carried out using PANalytical X'pert Highscore software, version 1.0e. The B-H bond stretch of the $NaBH_4$ and $Zn(BH_4)_2$ were compared via Perkin-Elmer *One* FTIR spectrometer and Renishaw 1000 Micro-Raman spectrometer.

2.3. Gravimetric and volumetric measurements

The thermal gravimetric analysis pertaining to the weight loss and the heat flow values for the reaction enthalpy during thermal decomposition of $Zn(BH_4)_2$ are estimated by using the TA instrument's DSC-Q10P and SDT-Q600 analytical tools. The DSC is calibrated with the certified Indium metal for the baseline and cell constant corrections. The calibration of TGA is performed in four steps with empty pan and standard sapphire disc. The ramp rate of 5° C/min was used for all the measurements. The isothermal volumetric sorption measurements have been carried out by Hy-Energy's PCT equipment. The volume calibration without and with sample is performed at a constant temperature with accuracy of ±1° C. The dehydrogenation and rehydrogenation parameters such as pressure and temperature have been optimized for the efficient hydrogen storage characteristics.

3. Results and Discussion

3.1. Powder x-ray diffraction

Figure 1 represents the powder x-ray diffraction pattern of the complex mixture ($2NaBH_4+ZnCl_2$) mechanically milled under N_2 for different time durations (30 min, 2h, 4h and 8h). It is clearly seen from the diffraction profiles that the NaCl peak and small satellite peaks corresponding to Zn-complex hydride structure present in the material irrelevant of the milling durations. Also the reaction completes within 30 minutes of milling where no initial peaks for $NaBH_4$ or $ZnCl_2$ peaks appears in the XRD profiles. The FWHM of the maximum intensity peaks increases with increasing ball milling duration say 30 mins to 8h.

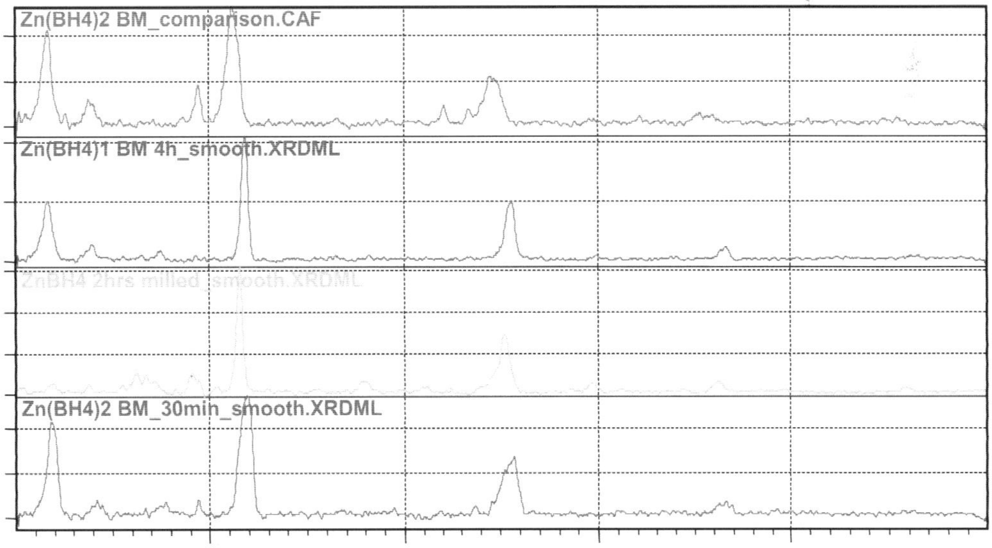

Figure 1 Powder x-ray diffraction pattern of ($2NaBH_4+ZnCl_2$) mixture milled unde N_2 ambient for different time durations (say 30 min, 2h, 4h and 8h).

3.2. FTIR and Micro-Raman spectra

The B-H bonding environment of $NaBH_4$ and $Zn(BH_4)_2$ are determined by FTIR and Raman spectroscopic techniques as shown in Figures 2 and 3 respectively. The FTIR spectra of BH_4^- ion in $NaBH_4$ has a characteristic band at 2283 and 2217 cm^{-1}, whereas in $NaBH_4$+$ZnCl_2$ ball milled mixture, it occurs at around 2084 cm^{-1} (bridging B-H bonds) as well as 2453 cm^{-1} (terminal B-H bonds). The Raman spectral shifts (figure 3) at these wavelengths confirm again the formation of $Zn(BH_4)_2$. The FTIR for the $NaBH_4$+$ZnCl_2$ ball milled and after hydrogen decomposition, shows different profile which corresponds to pure Zn and B according to the reaction:

$$Zn(BH_4)_2 \longrightarrow Zn + B + 4H_2 \qquad (3)$$

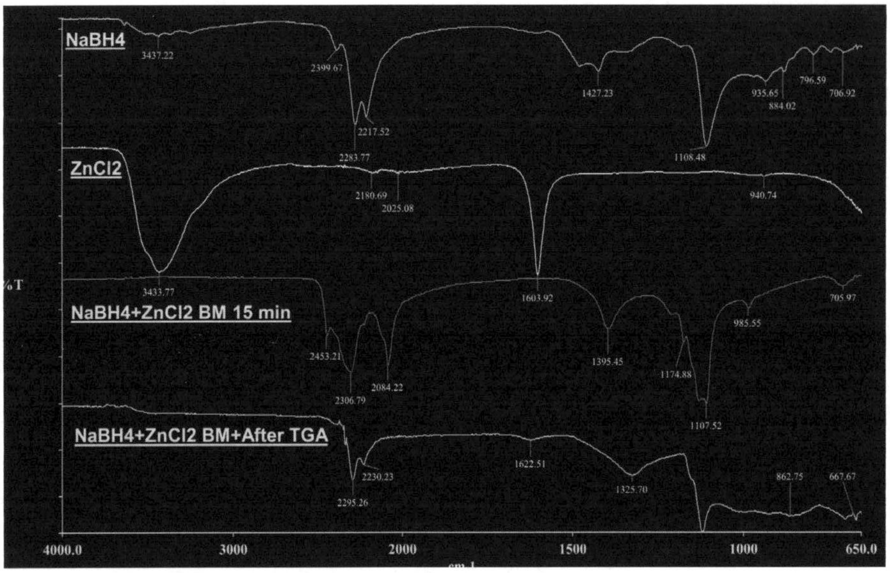

Figure 2 FTIR spectra of $NaBH_4$, $ZnCl_2$, $NaBH_4$+$ZnCl_2$ ball milled mixture before and after hydrogen decomposition.

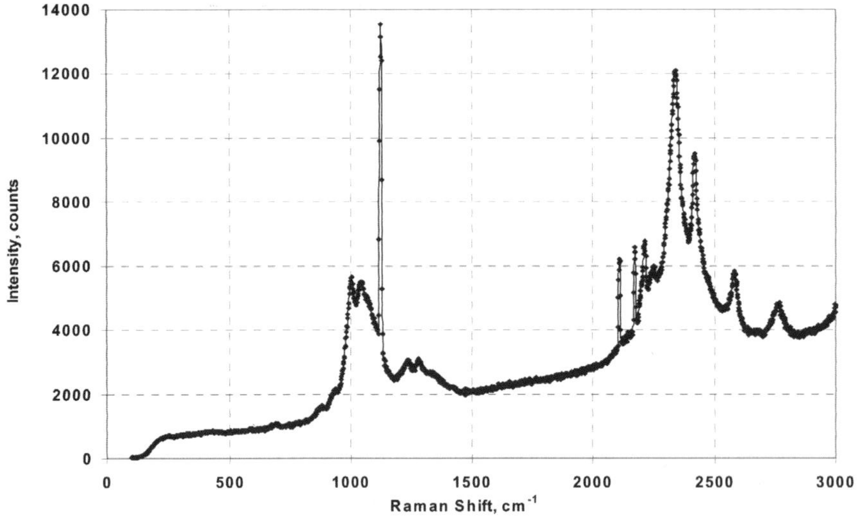

Figure 3 Micro-Raman spectrum of $NaBH_4$+$ZnCl_2$ milled mixture

3.3. Thermogravimetric Analysis

The gravimetric weight loss due to hydrogen decomposition from $NaBH_4$, $LiBH_4$ and $Zn(BH_4)_2$ are represented in Figure 4. It is known that $NaBH_4$ undergoes hydrolysis reaction to release hydrogen and hence the thermal decomposition process of $NaBH_4$ shows 0% weight loss even at temperatures >400° C. On the other hand, 10 -12 wt.% of H_2 decomposes at above 350° C for the case of $LiBH_4$ complex hydride. This is well agree with the theoretical prediction [12] according to the reaction,

$$LiBH_4 \longrightarrow LiH + B + 3/2\, H_2 \quad (4)$$

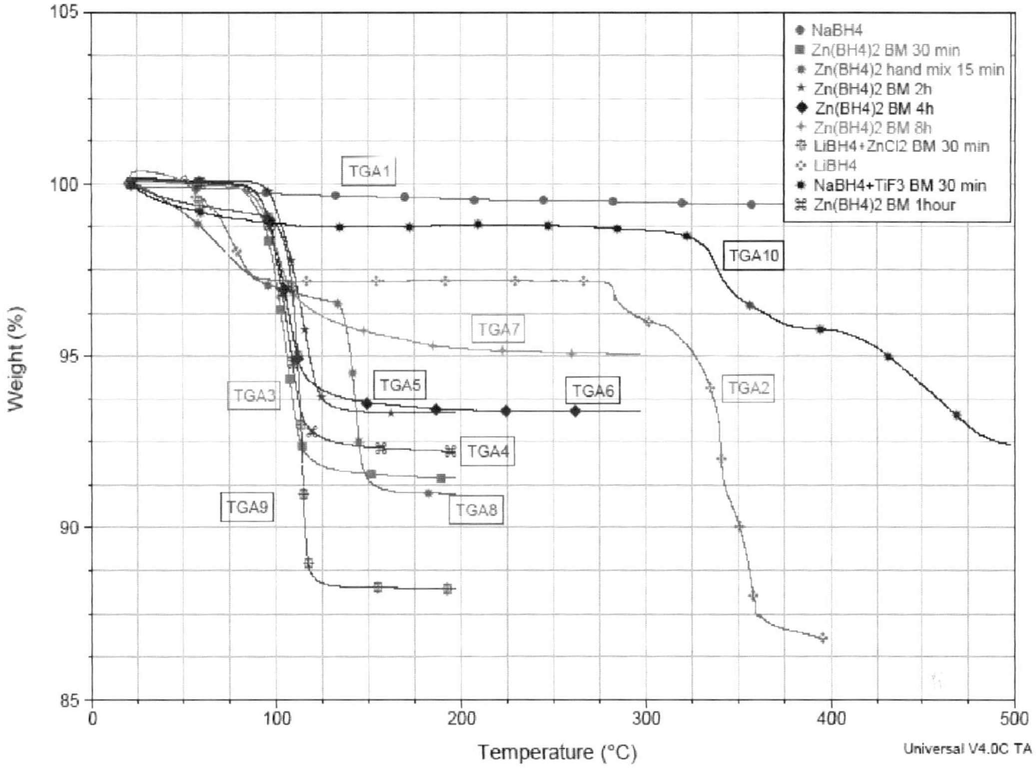

Figure 4. Thermogravimetric (weight loss analysis) profiles of (TGA1) $NaBH_4$, (TGA2) $LiBH_4$, $NaBH_4/ZnCl_2$ mechanically milled under vacuum, (TGA3) 30 min, (TGA4) 1h, (TGA5) 2h, (TGA6) 4h, (TGA7) 8h, (TGA8) hand mix under N_2 for 15 min, (TGA9) $LiBH_4/ZnCl_2$ milled for 30 min and (TGA10) $NaBH_4/TiF_3$ milled for 30 min.

The new complex borohydride, $Zn(BH_4)_2$ as obtained from the precursor complex hydride, $NaBH_4$ exhibits an excellent thermal decomposition of hydrogen with storage capacity of ~8 wt.% at around ~120° C. Also it is to be noted that the ball milling duration of 30 minutes yields optimum capacity and it decreases with increasing milling durations (say 1h, 2h, 4h and 8h). In order to demonstrate the effect of mechanochemical process, a parallel experiment have been carried out to hand grind the mixture of $ZnCl_2$ and $NaBH_4$ (1:2 mole ratio) under inert ambient for 15 minutes. It reveals from Figure 1, the hand mixed sample possesses at least 5.53 wt.% at slightly higher temperature, say 150° C. This is because of partial H_2 decpomposition from $Zn(BH_4)_2$ during the milling process and also the influence of O_2/moisture pickup (ppm level). The $Zn(BH_4)_2$ synthesized from $LiBH_4$ precursor shows the thermal decomposition of 10.84 wt.% at around 120° C. At the same milling duration of 30 minutes, performance of $LiBH_4$ as a precursor hydride is superior to the $NaBH_4$. This can be visualized from Figure 4. In order to compare the effect of Zn^{2+} on the $NaBH_4$, stoicheometric amount of (TiF_3+$NaBH_4$) have been mechanically milled for 30 minutes. The TGA profile of Ti^{3+} modified $NaBH_4$ exhibits T_{dec} of hydrogen at an elevated temperature above 350° C with total storage capacity of 5.3 wt.%. The weight loss due to H_2 docomposition and the milling duration for various precursors is given in Table 1.

Table 1. Thermogravimetric analysis of the complex borohydrides treated with various precursors and milling duration (refer also the Figure 4)

Sample Code	Sample name	Weight loss (%)	Weight loss (mg)	Operating temperature (°C)
TGA1	$NaBH_4$	0	0	400.00
TGA2	$LiBH_4$	10.21	1.408	346.76
TGA3	$Zn(BH_4)_2$; BM 30 min	8.456	1.640	113.36
TGA4	$Zn(BH_4)_2$; BM 1h	7.648	1.491	114.88
TGA5	$Zn(BH_4)_2$; BM 2h	6.681	1.485	122.82
TGA6	$Zn(BH_4)_2$; BM 4h	6.405	0.7607	114.73
TGA7	$Zn(BH_4)_2$; BM 8h	4.754	0.6720	112.74
TGA8	$Zn(BH_4)_2$ hand mix, 15 min	5.558	0.7817	146.63
TGA9	$LiBH_4/ZnCl_2$; BM 30 min	10.82	1.299	116.22
TGA10	$NaBH_4/TiF_3$; BM 30 min	2.803 3.336	0.1791 0.2132	351.39 478.86

3.4. Differential scanning calorimetry

The high pressure differential scanning calorimetry of the $NaBH_4$, $LiBH_4$ and $Zn(BH_4)_2$ mixtures ball milled at different time durations have been carried out and the DSC profiles are shown in Figure 5. The mixture $NaBH_4$+$ZnCl_2$ mixture milled for different time durations exhibits an endothermic peak due to hydrogen decomposition starting from 80° C and the enthalpies of decomposition for these materials are tabulated in Table 2. The thermal decomposition characteristics of $Zn(BH_4)_2$ exhibits an endothermic reaction process, with the ΔH value corresponds to -1 to -5 kJ/mol H_2 for the $NaBH_4$ precursor and -28 kJ/mol H_2 for the $LiBH_4$ precursor.

3.5 Summary

New complex hydrides, $Zn(BH_4)_2$ have been synthesized by an inexpensive mechanochemical process involving ball milling of stoicheometry of $NaBH_4$ and $ZnCl_2$ under inert ambient. Extensive characterization such as structural, bonding environment, thermal and volumetric measurements are done by powder x-ray diffractometry, FTIR, Raman spectrometry, DSC, TGA and hydrogen sorption tools. $Zn(BH_4)_2$ possesses high hydrogen storage capacity of ~8.0 wt.% under thermal decomposition temperature of 100° C. It also exhibits reversibility of dehydrogenation and rehydrogenation characteristics under moderate temperature and pressure conditions. The stability of the structure due to strong B-H interaction inhibits the reversibility of reaction at low temperatures. Right now the research direction is focused on destabilizing the structure by catalytic doping or lattice substitution mechanisms.

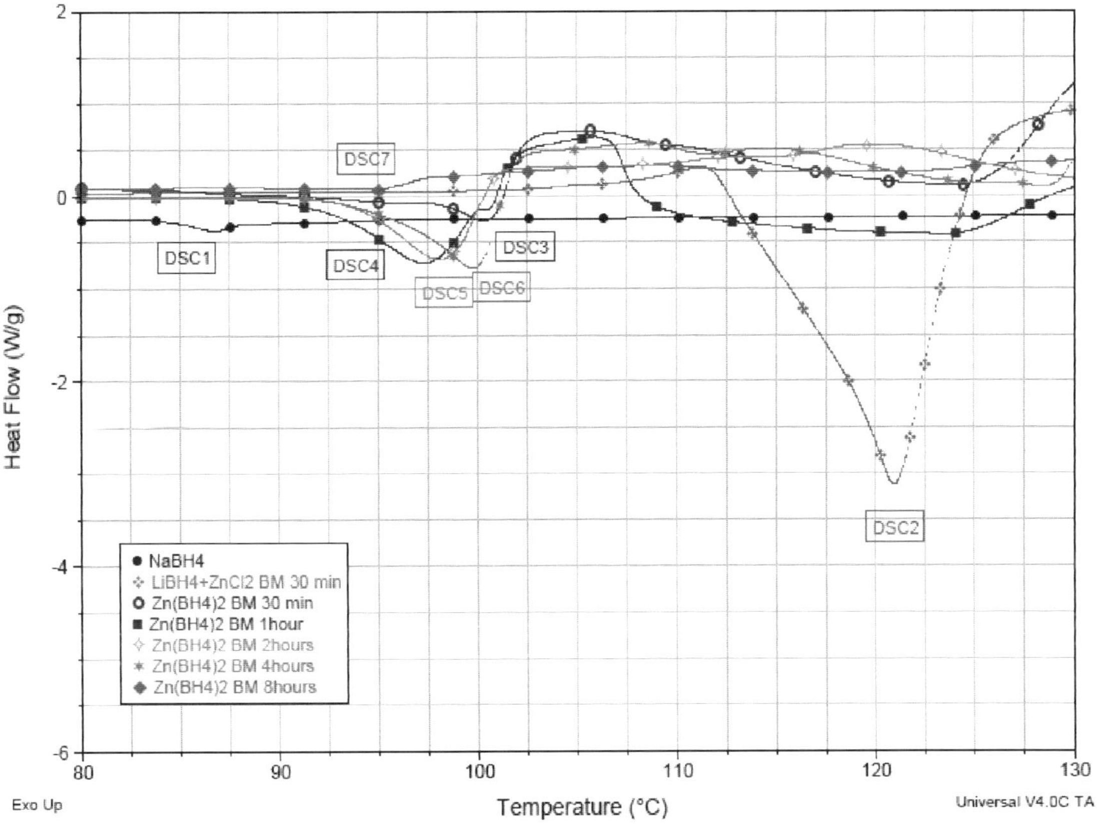

Figure 5. Heat flow Differential scanning calorimetric curves of (DSC1) $NaBH_4$, (DSC2) $LiBH_4/ZnCl_2$ mechanically milled under vacuum 30 min, $NaBH_4/ZnCl_2$ mechanically milled under vacuum for (DSC3) 30 min, (DSC4) 1h, (DSC5) 2h, (DSC6) 4h, (DSC7) 8h

Table 2. The enthalpy of decomposition of hydrogen (kJ mol^{-1} H_2) from the complex borohydrides via Differential scanning calorimetric measurements

Sample code	Sample Name	ΔH_{dec} (kJ mol^{-1} H_2)	Onset temperature (°C)	Peak temperature (°C)
DSC1	$NaBH_4$	-0.22	84.52	86.57
DSC2	$LiBH_4$+$ZnCl_2$; BM 30min	-28.3	115.06	120.98
DSC3	$Zn(BH_4)_2$ BM 30 min	-1.225	98.37	11.99
DSC4	$Zn(BH_4)_2$ BM 1h	-3.55	92.20	97.17
DSC5	$Zn(BH_4)_2$ BM 2h	-4.54	93.80	98.36
DSC6	$Zn(BH_4)_2$ BM 4h	-5.4	94.88	99.94
DSC7	$Zn(BH_4)_2$ BM 8h	-0.58	90.11	95.23

Acknowledgement

Authors wish to acknowledge the financial support from US Department of Energy (Project code: DE-FG36-04G014224) and University of South Florida for carrying out this work.

References and Notes

[1] L. Schlapbach, A. Zuttel, Nature, 414, 2001, 353;
[2] B. Bogdanovic, M Schwikardi, J. Alloys Compd., 253-254, 1997, 1-9;
[3] http://www.eere.energy.gov/hydrogenandfuelcells/posture_plan04.html
[4] Crabtree. G.W.; Dresselhaus, M.S.; Buchanan, M.V. *Physics* Today **2004,** 39-44.
[5] Schlapbach, L.; Zuttel, A.*Nature* **2001,** 414, 353.
[6] Bogdanovic, B.; Schwikardi, M. *International Patent* **1997,** WO97/03919.
[7] Jensen, C.M.; Zidan, R. *US Patent* **2002,** 6,471,935 B2.
[8] http://www.eere.energy.gov/hydrogenandfuelcells/posture_plan04.html
[9]http://www.eere.energy.gov/hydrogenandfuelcells/pdfs/freedomcar_targets_explanations.pdf
[10] http://hydpark.ca.sandia.gov
[11] Li, Z.P.; Liu, B.H.; Arai, K.; Morigazaki, N.; Suda, S. *J. Alloys Compd.* **2003,** 356 357, 469-474.
[12] Zuttel, A.; Wenger, P; Rentsch, S.; Sudan, P.; Mauron, P.; Emmenegger, C.; *J. Power Sources* **2003,** 118, 1.
[13] Au, Ming. *Material Research Society fall meeting, Boston, MA* **2005.**
[14] William, D.; Davis, L.; Mason, S.; Strgeman, G. *J. Am. Chem. Soc.* **1949,** 71, 2775-2781.
[15] Lodziana, Z.; Vegge, T. *Phys. Rev. Lett.* **2004,** 93, 14, 145501-1.
[16] Frankcombe, T.J.; Kroes, G-J.; Zuttel, A. *Chem. Phys. Lett.* **2005,** 405, 73-78.
[17] Grochala, W.; Edwards, P.P. *Chem. Rev.* **2004,** 104, 1283-1315.
[18] Narashimhan, S.; Balakumar, R. *Aldrichimica Acta* **1998,** 31, 1, 19-26.
[19] Edwards, P.; Grochala, W.; Book, D.; Harris, I.R. *PCT Int. Appl.* **2004,** 21.
[20] Eun Jeon, Young W. Cho, *J. Alloys Compd.* **2006,** Article in press.

Fundamental Structural Properties of Single-Walled Carbon Nanotubes: Reopening the Debate over Hydrogen Storage in Nanotubes

Sandeep Agnihotri and Yun Fat-Lam*
Department of Civil and Environmental Engineering, University of Tennessee, Knoxville

*Corresponding author

Introduction

We have developed a molecular simulation and experimental adsorption based methodology to determine, for the first time, the fraction of open-ended single-walled nanotubes (SWNTs) in several purified samples (Agnihotri et. al., *Langmuir* 2005, 21(3):896-904).[1] The approach and results are presented later in this paper. We have discovered, also for the first time, that purified samples of nanotubes could have as little as only 45% nanotubes that are open-ended or unblocked for an unrestricted adsorption; i.e., approximately half fraction of samples was found redundant for adsorption. The maximum adsorption capacity and true potential of nanotubes to store hydrogen, therefore, would be much greater if adsorption is normalized by this new information on structural properties of nanotubes or if techniques to open and characterize all nanotubes can be developed. This discovery has stirred the debate over hydrogen adsorption capacity of carbon nanotubes and the use of nanomaterials in general for efficient energy storage. The proposed study is relevant to scientists and economists all over the world due to its direct implications on the environment and the global economy, and its potential to answer the decisive question: Can nanotubes ever be used to store energy?

Carbon nanotubes are manufactured nanomaterials. They have a well-defined atomic structure of carbon atoms, and are typically 10,000 to 100,000 times thinner than human hair.[2] The unique 1D nanometric structure of single-walled carbon nanotubes, in particular, imparts remarkable properties.[2] The hollow nanoscaled cylindrical structure of SWNTs is especially attractive for their use as adsorbents[3-10] and to store energy efficient gases, such as H_2.[11-17] However, in spite of a tremendous effort to better understand adsorption on nanotubes, an application utilizing nanotubes as an adsorbent is yet to be realized. Furthermore, the issue of adsorption surface area of nanotubes, as measured by gas adsorption, still remains unresolved. The reported BET surface area of SWNTs ranges anywhere from 150 m^2/g to 1,587 m^2/g, depending on sample purity and morphology.[3-13] It is equivalent to that of mesoporous activated carbons, which are at least a 1,000 times cheaper than nanotubes. Moreover, average H_2 storage capacity of nanotubes is ≈ 4.2 wt% at 25 °C and 100 atm;[18,19] although adsorption capacities as high as 8 wt% at similar conditions have also been reported.[20] The DOE goal for H_2 storage in a material is at least 6.5 wt%. Scientists, therefore, have vehemently disputed over the issue of H_2 storage in nanotubes[21] and the use of nanotubes as an adsorbent for energy storage still remains an open debate.

Fig.1. Adsorption sites on a homogeneous bundle of SWNTs: (1) internal, (2) interstitial channels, (3) external grooves, and (4) external surface.

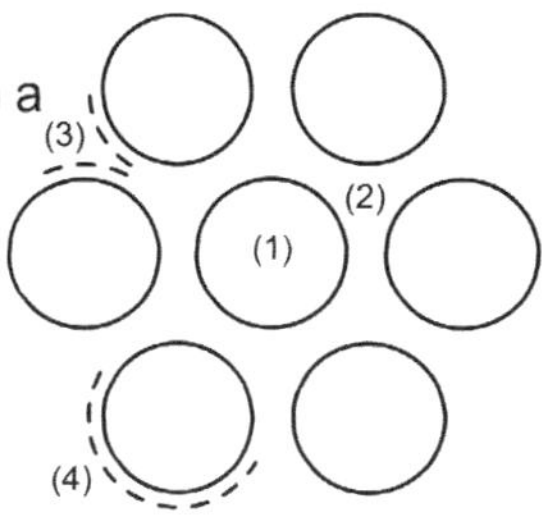

The present knowledge of adsorption, and thus energy storage, in nanotubes is grossly undermined. This is because a structural property that is fundamental but extremely relevant to adsorption in nanotubes has not been addressed until now. It is the *fraction of open nanotubes* in a sample. Only open-ended nanotubes can utilize the hollow interior volume of the nanotubes. Typically, nanotubes are closed on either or both ends, but they can be opened by oxidative processes.[22] Only open-ended SWNTs facilitate adsorption inside nanotubes, and thus, utilize the very porosity that makes them a suitable candidate for adsorption applications. Due to strong van der Waals forces, SWNTs adhere to each other and form ropes or bundles, which have four possible sites for adsorption (Figure 1).[3,5] Site **1** provides a large volume for adsorption and gas storage, but it can be accessed only for those nanotubes that are open-ended. Therefore, the true potential of nanotubes to store energy can only be realized if the fraction of open-ended nanotubes is known.

Experimental and Simulation Methods

The study is accomplished by comparing results from adsorption tests obtained by experiments with those from grand canonical Monte Carlo (GCMC) simulation. The study is performed on *purified* samples of SWNTs to ensure that the results reflect properties of nanotubes, and not those of impurities that are present in typical samples. This is an important aspect of our research, as most previous studies[10-17] were performed on samples that contained uncharacterized amounts of impurities. A purified samples of SWNTs was characterized for (i) purity by transmission electron microscopy, (ii) nanotube diameter distribution by Raman scattering experiments, and (iii) N_2 adsorption isotherms, BET surface area and pore volumes by standard N_2 adsorption (77 K) techniques. The information gathered about the sample morphology is used to develop an adsorbent model for SWNTs to be used for GCMC simulations of N_2 adsorption (77 K). The simulation results will be compared with experimental adsorption results for interpretation. Following is the description of the simulation procedure.

Simulation Procedure: In order to model adsorption of samples, a simulation box that would contain a heterogeneous distribution of SWNTs would be the most appropriate structural arrangement of nanotubes. However, making heterogeneous nanotube bundles[7] that closely resembles the bundle size and distribution of an experimental sample could be quite complicated, because of which an alternative methodology is invented. Here GCMC simulation of N_2 adsorption (77 K) were performed on homogenous arrays of SWNTs of diameters the same as those in experimental samples, as obtained from Raman scattering experiments. The position of individual carbon atoms in a nanotube was assumed unimportant at the temperatures of interest in this study, i.e., each nanotube in the bundle was considered to be a smooth structureless nanocylinder. Total adsorption on a bundle was calculated in two steps; external adsorption: adsorption on the outer surface of the nanotube bundles (i.e., sites **3** and **4**, Fig.2a), and internal adsorption: adsorption inside the bundles (i.e., sites **1** and **2**, Fig.2b). The

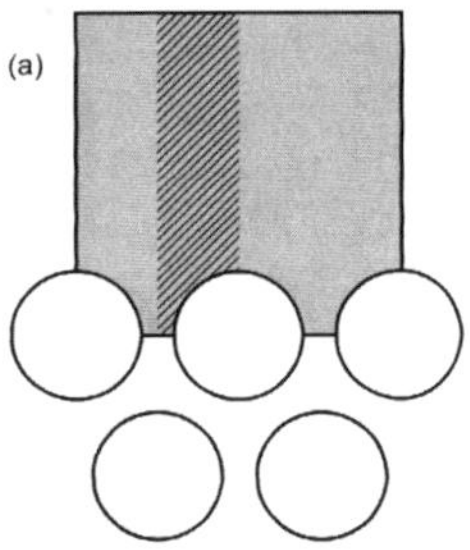

Fig.2. Cross section of unit simulation box for GCMC study of nitrogen adsorption onto different adsorption sites of a bundle of open-ended SWNTs: (a) external groove sites and external surfaces, (b) the internal (endohedral) and interstitial channels. The shaded area represents the effective volume probed during the simulation.

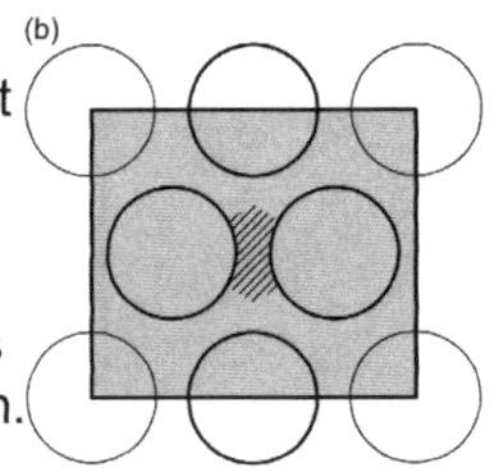

intertube distance for all simulations is kept fixed at 3.4 Å to imitate SWNTs adhering to each other by van der Waals interaction which forms bundles.

N_2 molecules were treated as structureless spherical particles which interact via dispersive forces only. The interaction between N_2 molecules is modeled by the 12-6 Lennard-Jones (LJ) potential $u_{ij}(r) = 4\varepsilon_{ij}\left[(\sigma_{ij}/r)^{12} - (\sigma_{ij}/r)^{6}\right]$, where *r* is the intermolecular distance, as is the interaction between the carbon atoms of a nanotube and each N_2 molecule. The cross terms were obtained using the standard Lorenz-Berthelot combining rules: $\varepsilon_{ij} = \sqrt{\varepsilon_i \varepsilon_j}$ and $\sigma_{ij} = (\sigma_i + \sigma_j)/2$. The well depths $\varepsilon_{C\text{-}C}/k_B$ and $\varepsilon_{N\text{-}N}/k_B$ used in calculations are 28.0K and 100.4K, and collision diameters $\sigma_{C\text{-}C}$ and $\sigma_{N\text{-}N}$ are 3.4Å and 3.609Å, respectively. Once the external and internal adsorption capacities of homogenous bundles are determined by simulation, the total adsorption, $q_{sim}(P/P_o)$, for a sample was calculated by averaging the adsorption contribution from bundles of nanotube diameters relevant to that sample:

$$q_{sim}(P/P_o) = S_P\, q^{s}_{sim}(P/P_o) + \alpha(1-\eta) q^{i}_{sim}(P/P_o)$$

where, $\left\{q^{s}_{sim}, q^{i}_{sim}\right\} = \sum_{D=sample} w_D \left\{q^{s}_{sim,D}, q^{i}_{sim,D}\right\}(P/P_o)$, S_P is the external surface area of nanotube bundles in the sample (m^2/g), and $q^{s}_{sim,D}(P/P_o)$ and $q^{i}_{sim,D}(P/P_o)$ are adsorption capacities for the external surface (s) and internal volume (i) of a homogeneous bundle of nanotube diameter, *D* at relative vapor pressure P/P_o, 100η is the wt% of impurities in the sample and w_D is the weight fraction of nanotubes of diameter, *D*, and *α* is a scaling parameter that adjusts an entire internal adsorption isotherm to yield a total isotherm which fits the experimental isotherm. S_p and *α* are structural parameters of samples, which are determined by plotting experimental and simulated N_2 isotherms. The parameter *α* is termed as the *fraction of open-ended SWNTs* in the sample.

Results and Discussion

The study was conduced on two commercially available and purified samples of single-walled carbon nanotubes (SWNTs). The samples, labeled EA95 and CVD80 where the numeral indicates the wt% of nanotubes in the sample, were produced by electric arc (EA) and HiPco™ chemical vapor deposition (CVD) processes that are known to produce nanotubes of different morphologies. Raman scattering experiments showed that majority of SWNTs in sample EA95 were 1.52 nm wide (other sizes: 1.11 and 1.4 nm) and those in sample CVD80 were 0.9 nm in diameter (other sizes: 1.02, 1.07, 1.17 and 1.14 nm). The samples were characterized by standard N_2 adsorption at 77 K to determine isotherms, BET surface area and porosity.

Figure 3 compares the experimental and simulated N_2 (77 K) isotherms on sample EA95. It clearly shows a near perfect replication of experimental isotherms by assuming 45% open nanotubes in the test sample. Therefore, only 45% of total internal volume of nanotubes was available. Furthermore, the BET surface area would increase from 500 m^2/g to 1,200 m^2/g if all SWNTs were open, and the total pore volume and adsorption capacities of the sample would also increase by 125%.

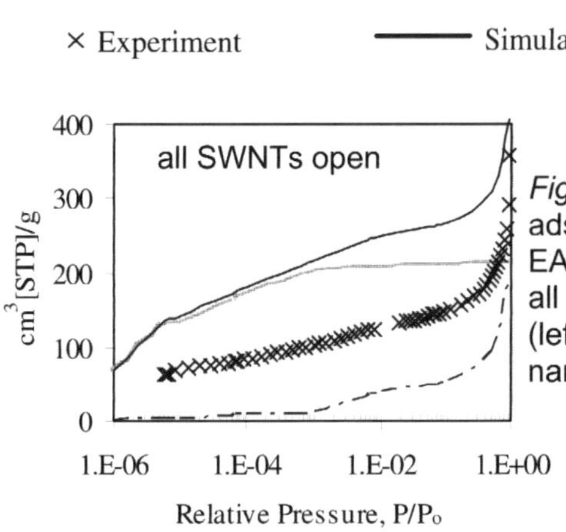

Fig.3. Simulated N_2 adsorption isotherms of EA95 sample assuming all nanotubes open (left), and 45% nanotubes open (right).

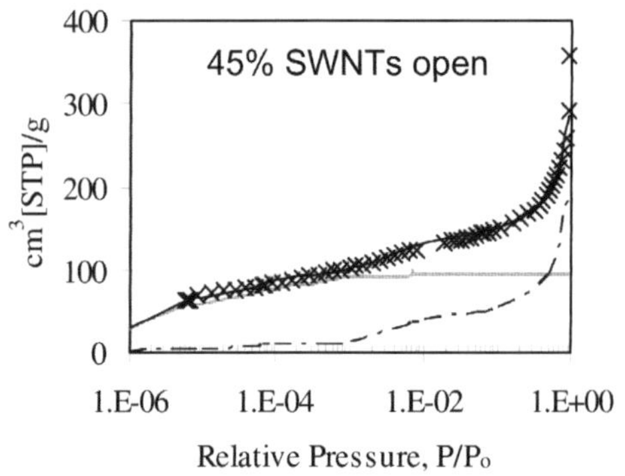

Mechanisms of N_2 adsorption in SWNTs were explored by grand canonical Monte Carlo simulations (Figure 4). It was found that (i) adsorption on the peripheral surface of the bundles is at least as significant as that inside of the nanotubes, (ii) external adsorption initiates on the groove sites at $P/P_o \approx 10^{-6}$, and once grooves are saturated only then N_2 adsorbs onto the rest of the external surface, (iii) external adsorption is insensitive to nanotubes size, (iv) adsorption in the interstitial channels between nanotubes does not occur in the homogenous arrays unless the nanotubes are wider than 1.4 nm, and (v) samples comprising thin nanotubes (say, 0.9 nm) adsorb more on the external surface of the bundles and less within the nanotubes as compared to samples with wide nanotubes (e.g. 1.52 nm diameter).

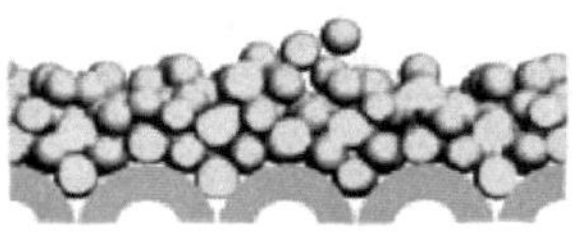

Fig.4. Snapshots of N_2 molecules (77 K) adsorbing on the external surface (left) and inside (right) homogeneous bundles of SWNT of 9 Å diameter ($P/P_o = 0.1$).

In summary, our nanoscale simulations were able to accurately predict the macroscale phenomenon of adsorption as observed via experimentation on gram level quantities of samples. The results illustrated the accuracy of the overall methodology invented for this study, i.e., simplifying heterogeneous bundles in experimental samples as homogenous bundles of diameter distribution obtained from a technique familiar to most nanotube scientists and then calculating adsorption outside and inside the bundles.

Impacts and Relevance

This study reveals new scientific information on fundamental physical and chemical properties of carbon nanotubes. It will advance science of nanotubes and contribute greatly to our current understanding of these nanomaterials. The results from this study will stimulate other interesting researches in this area. We envision several studies that stem from the proposed research, such as studies related to synthesizing and commercializing functional carbon nanomaterials for adsorption and transport of fluids. This study will also lead to development of techniques to *open* more nanotubes so that their entire internal pore volume could be utilized. Such techniques are imperative to explore the possibility of using nanotubes for other adsorption related applications. This would create a number of opportunities for stimulating research related to adsorption, storage and transport of gases through carbon nanotubes. For example, controlled flow of fluids through channels of *open* ended nanotubes can be accomplished by incorporating electron sensitive functional groups on the open ends which can then be manipulated by an external source of energy to allow or stop the flow of a fluid. Such structures and other related structures, such as nanotube gas cylinders,[4] nanotube channels,[24] and membranes[25] for transporting fluids and drug delivery have been synthesized in the recent past; however, the issue of homogeneity of transport properties throughout a sample and replication of samples with similar transport properties have been stalled by the very fact that there is little knowledge and no control over the amount of nanotubes that are open and able to facilitate fluid transport. There are several reports of advances in nanotube manufacturing with predictions of ton-level quantities to be produced daily in the near future. The scientific information generated from this research will greatly enhance fundamental and applied research of these nanomaterials.

References

1. Agnihotri S, Mota PB, Rostam-Abadi M, Rood M.J. *Langmuir* 2005, 21(3):896-904.
2. Dresselhaus MS *Topics in Applied Physics* 80. Springer 2000.
3. Agnihotri S, Rostam-Abadi M, Rood MJ. *Carbon* 2004, 42 (12-13):2699-2710.
4. Matranga C, Bradley B. *Journal of Physical Chemistry B* 2005; 109(19):9209-9215.
5. Zhao J, Buldum A, Han J, Lu JP. *Nanotechnology* 2002; 13(2):195-200.
6. Chang H, Lee JD, Lee SM, Lee YH. *Applied Physics Letters* 2001; 79(23):3863-3865.
7. Shi W, Johnson JK. *Physical Review Letter* 2003, 91(1):015504-4.
8. Varghese OK, Kichambre PD, et al. *Sensors and Actuators B* 2001; 81(1):32-41.
9. Talapatra S, Zambano AZ, et al. *Physical Review Letters* 2000; 85(1):138-141.
10. Cinke M, Li J, Chen B, Cassell A, et al. *Chemical Physics Letters* 2002; 365:69-74.
11. Zhu H, Cao A, Li X, Xu C, et al. *Applied Surface Science* 2001; 178(1):50-55.
12. Yin YF, Mays T, McEnaney B. *Langmuir* 2000; 16(26):10521-10527.
13. Nishimiya N, Ishigaki K, et al. *Journal of Alloys and Compounds* 2002; 339(1):275-282.
14. Ma Y, Xia Y, Zhao M, Ying M. *Physical Review B* 2002; 65(15):155430.
15. Darkrim F, Levesque D. *Journal of Chemical Physics* 1998; 109(12):4981-4984.
16. Levesque D, Gicquel A, et al. *Journal of Physics- Condensed Matter* 2002; 14(40):9285.
17. Cracknell RF. *Molecular Physics* 2002; 100(13):2079-2086.
18. Liu C, Fan YY, Liu M, Cong HT, et al. *Science* 1999; 286(5442):1127-1129.
19. Liu C, Yang QH, Tong Y, Cong HT, et al. *Applied Physics Letters* 2002; 80(13):2389-2391.
20. Heben MJ, Dillon AC, Gannett T, et al. *MRS symposia proceedings* 2001; 633: A9.1.
21. C& EN News: 80 (# 2), pp. 25-28. January 14, 2002.
22. Fujiwara A, Ishji K, et al. *Chemical Physics Letters* 2001; 336(3):205-211.
23. Martin GM, Siepmann JI. *Journal of Physical Chemistry B* 1998. 102:2569-2577.
24. Huang BD, Xia YY, Zhao MW, et al. *Journal of Chemical Physics* 2005, 122(8):084708.
25. Nednoor P, Chopra N, Gavalas et al. *Chemistry of Materials* 2005, 17(14): 3595-3599.

Safe, Efficient Storage of Hydrogen in Liquid Ammonia and Ammonia-Based Solid Chemicals

John H. Holbrook, Energy Science & Technology Directorate, Pacific Northwest National Lab, Richland, WA 99352, Norman G. Olson, BECON, Iowa Energy Center, Des Moines, IA 50010, Robert K. Graupner, Oregon Sustainable Energy LLC, Hillsboro, OR 97124, and R. Doyle Miller, Intelligent Energy—Mesofuels, Albuquerque, NM 87113.

Abstract—Pure anhydrous ammonia (NH3) provides an excellent storage medium for hydrogen, and two known solid organic chemicals related to ammonia, urea and guanidine, are also good storage media. Anhydrous ammonia is about 18 percent hydrogen by weight percent and has all of the advantages associated with a liquid fuel—high volumetric density, easy to store, easy to pump, easy to transport, and easy to deliver. It also has an existing and proven distribution infrastructure and is one of the highest tonnage chemicals produced in the world. Furthermore, gaseous hydrogen can be readily produced (reformed) from anhydrous ammonia at the highest efficiency of any hydrogen-containing chemical. On the other hand, it is well known that ammonia is toxic which results in some handling/delivery safety concerns. We will discuss safe handling of anhydrous ammonia. It turns out as well that both urea and guanidine (solids at standard temperature and pressure) have essentially zero toxicity, are not explosive, and do not exhibit any other public or environmental hazard. Both are high in hydrogen content via their ammonia content, and both can be reacted with water under appropriate conditions to provide ammonia, which can then be reformed to hydrogen as necessary. We will discuss the features of these two chemicals and their potential applications as safe, energy-dense hydrogen energy carriers.

INTRODUCTION

Hydrogen is a very attractive energy medium for the near-term and for the future. Hydrogen is relatively easy to produce, can free us from dependence on oil-based fuels, and is the only chemical (other than its cousin, ammonia) that burns carbon-dioxide, and for practical purposes, NOx free. This is, of course, because hydrogen contains no carbon in its chemical make-up, compared to carbon-based fossil fuels. And, with the potential for delivering clean energy through fuel cells and internal combustion engines with essentially zero environmental impact, hydrogen is the model fuel of the future.

The pivotal problem with hydrogen as a fuel, of course, is its very low volumetric net energy density, either in the form of a compressed gas or as a cryogenic liquid. Esssentially, the problem is that if you need to move hydrogen from its point of production to its point of use, this can consume a sizable fraction of the energy contained in the hydrogen itself. And, in turn, this can compromise wide-scale use of hydrogen, either as liquid or gas, as an efficient transporter of energy.

This paper will address hydrogen energy storage and delivery via a "carrier", an economical and plentiful industrial chemical—ammonia. Ammonia is developing considerable support as an excellent "one-way" hydrogen carrier. This is largely because ammonia is nearly 18 percent hydrogen by weight, is a pumpable liquid at moderate pressures and ambient temperatures, and has an established and proven delivery infrastructure in the U.S., and is relatively straightforward to liberate the hydrogen from the ammonia at the point of use.

STORAGE/DELIVERY OF HYDROGEN BY AMMONIA

To provide sufficient background, here are some important and relevant points--

Over 100 million tons a year of anhydrous ammonia (NH_3, 18% hydrogen) are produced currently (2005) annually world-wide. Compare this to the 40 million tons of annual hydrogen production forecasted for a mature "hydrogen economy" by DOE.

Over 2000 miles of ammonia pipeline are currently operating in the continental U.S., as a means of delivering fertilizer to farmers.

Ammonia is routinely and safely shipped in large quantities via train, truck, barge, pipeline and trans-oceanic ship.

Ammonia is the easiest chemical to reform and re-produce hydrogen, say compared to methane, methanol, and ethanol, etc.

Aside as a hydrogen carrier, ammonia can be an important direct fuel—powering fuel cells and internal combustion engines without intermediate reforming to hydrogen.

Because of the widespread production/consumption of ammonia for agricultural purposes, the price of ammonia fuel can be expected to be stabilized and competitive.

Liquid anhydrous ammonia (NH_3)is more than 50% more energy dense than liquid hydrogen, and over 3 times the energy density, by volume, as 10, 000 psi hydrogen gas.

Ammonia is not especially corrosive and can be safely stored and shipped in vessels nd piping made of mild steel, or even aluminum alloys.

The primary drawback of ammonia as a fuel is its toxicity.

Small ammonia storage tanks can be safely used as fuel tanks for vehicles.

Statistics show that significant releases of gasoline cause fatality rates five times higher than significant releases of ammonia.

Ammonia is much lighter than air and dissipates rapidly when leaked to the atmosphere

Ammonia storage and transport systems can meet 2015 USDOE targets today.

Ammonia is not, in itself, a greenhouse gas

LARGE-SCALE STORAGE/TRANSPORTATION COSTS

Refrigerating or liquefying hydrogen to -423 F (–253 degrees Celsius) uses the equivalent of 25% to 30% of its energy content, and requires special materials and handling. To cool one pound (0.45 kg) of hydrogen requires at least 5 kWh of electrical energy. Double walled tanks with a combination of expensive insulation systems and a vacuum are required between the two walls. Storage losses are a significant issue. Cryogenic (-423 F) hydrogen storage costs reported in a publication entitled, "Hydrogen Storage: State-of-the-Art and Future Perspective" ©2003, ranged from $20 to $4,500/kg (hydrogen) depending on storage capacity. The $20 per kg (hydrogen) corresponds to a storage size of ~ 20,000 metric tonne.

Costs for a complete 25,000 metric tonne low-temperature (-28F) ammonia storage facility (including refrigeration, loading and unloading systems) recently was bid at $6 million in Eastern, Iowa. The tanks are constructed of mild steel and insulated with approximately 12" of perlite insulation (or equivalent). A cost of approximately $1.61 per kg (hydrogen), more than an order of magnitude lower than the lowest cost listed for liquid hydrogen storage costs.

Per mile capital costs for 12 inch natural gas and hydrogen pipelines were estimated to be $600,000 and $1,000,000 respectively in a 2002 Argonne National Lab report entitled "Cost of Some Hydrogen Fuel Infrastructure Options".

The capital cost per mile for a 12 inch ammonia pipeline were calculated to be approximately $300,000 in a 1999 paper provided by an Iowa-based ammonia production facility. (Note: Steel prices nearly doubled from November 2003 to November 2004 but were fairly stable from 1999 to 2002.)

SMALL-SCALE, SAFE AMMONIA STORAGE

Before ammonia-based hydrogen generators will gain acceptance, the problem of safe ammonia storage must be addressed. Ammonia is a toxic gas that can rapidly damage the eyes and respiratory tract upon exposure to concentrations in the range of 500 to 1000 ppm. Even brief exposure to higher concentrations (>5000 ppm) can lead to respiratory failure and death.

Drawbacks of Conventional Storage Tanks

Targeted small-scale hydrogen generators currently under development employ lightweight storage tanks made from either aluminum or titanium . These tanks have a mass of approximately 120 g and an ammonia storage volume of 0.8 liters. The lightweight storage tanks (aluminum and titanium) are designed to withstand over 1000 psig to ensure they do not burst in response to ammonia vapor pressure, which can exceed 250 psi at credible ambient temperatures. However, these storage tanks are not designed to withstand bullets or puncture from sharp objects. Department of Transportation (DOT) approved ammonia storage tanks employ thick metal walls and are generally much heavier. Fig. 4b shows a DOT-approved tank with an ammonia capacity of 1.5 liters and an empty mass of 2.3 kg. As discussed earlier, anhydrous ammonia has a fuel-cell energy density of 2600 W-h/kg. If ammonia is stored in a 0.8-liter titanium tank, the effective energy density (including only the mass of ammonia and the tank) is reduced to 2000 W-h/kg. Use of the 1.5-liter, DOT-approved tank yields an effective energy density of 750 W-h/kg.

The safety concerns associated with storing liquid ammonia in thin-walled storage tanks led to the consideration of the use of semi-permeable, high-void-fraction monoliths as a safe storage media for liquid ammonia. Because the ammonia is stored as a liquid in high-void-fraction material, the storage density is high. But because the liquid ammonia is trapped within the pores of the monolith, a tank leak will not result in expulsion of liquid ammonia regardless of the leak location. The gas-phase ammonia can only escape at a slow rate, which is determined by the chemical and physical properties of the storage monolith. The liquid ammonia is stored inside a low-permeability, high-void-fraction, monolithic structure that adds minimal weight yet effectively prevents rapid release of ammonia. A rigid tank, composed of metal, fiberglass, or a suitable plastic, surrounds the ammonia-storage monolith. Ammonia is forced into the monolith thereby filling the void spaces with liquid. This approach offers very high storage density, compact size, and a controlled ammonia release rate even in the event of tank puncture. Rapid release of ammonia from the monolith is controlled by thermal and mass-transfer effects. To escape the monolith, the ammonia must permeate through the small pores to near the surface of the monolith.

STORAGE OF HYDROGEN IN AMMONIA-BASED SOLID CHEMICALS

Safe Ammonia Storage in Chemical Compounds

Ammonia can be safely stored as a chemically bound form in the compounds urea and guanidine. These compounds can be reacted with water at moderate temperatures to release the stored ammonia as shown by the following equations, where the numbers refer to the standard enthalapies of formation:

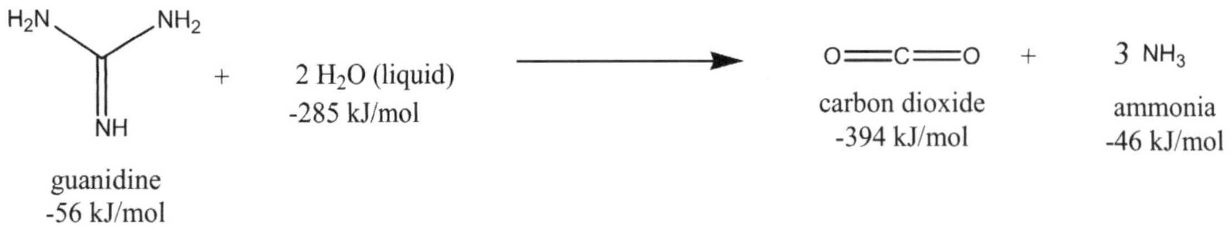

Urea is a common stable compound which is currently used as an ammonia source for the reduction of nitrogen oxides in electrical power plant exhaust gases. While easily handled it has a limited solubility in water, which reduces its effective energy density when distributed in liquid form. Urea produces two volumes of ammonia for each volume of carbon dioxide released. This mixture may be further diluted by steam from the excess water required to dissolve the urea prior to reaction. As a result, the product ammonia gas from urea decomposition may require purification prior to use as a fuel. In comparison, guanidine is highly soluble and can be dissolved in the amount of water which is required for its stoichiometric decomposition. Because guanidine has a low melting point (~ 50° C) it can also be stored on the vehicle in solid form and delivered to the reaction chamber in molten form. Negligible amounts of water vapor are released with the ammonia when combined with exactly the amount of water needed for decomposition. Then one volume of carbon dioxide is released for every 3 volumes of ammonia produced providing a high energy density mixture which can immediately be used as a fuel without a separation step. The low temperature heat (~ 100 °C) required to decompose a mole of guanidine into ammonia (94 kJ/mole) can be easily obtained from the engine exhaust because it represents only 20% of the exhausted heat assuming an engine efficiency of 50%. The water needed for decomposition can be captured from the vehicle exhaust eliminating the need to carry additional water onboard the vehicle. The combination of these two factors gives guanidine a very high energy density (4.7 kWhr/L using captured water and waste heat after allowing 20% for storage and processing equipment). When considering the weight of the required fuel storage tanks, guanidine has more than twice the available energy density of compressed hydrogen. Unlike hydrogen, guanidine would not require construction of an expensive delivery infrastructure. As a nonflammable and relatively safe material, guanidine could be distributed in recyclable containers sold at convenience and grocery stores in a manner similar to current sales of bottled propane. The use of a solid fuel will also simplify the design of vehicles because the fuel can be positioned to absorb the energy of collisions and protect the occupants from injury as opposed to requiring special shielding in the case of conventional highly flammable fuels.

256g Emissions from a Hydrogen-Compressed Natural Gas Fueled Ford F-150

Henry Ng[1], **Robert W. Peters**[2], and Fouad H. Fouad[2]. (1) Argonne National Laboratory, 9700 South Cass Avenue, Argonne, IL 60439, (2) University of Alabama at Birmingham, Department of Civil and Environmental Engineering, 1075 13th Street South, Birmingham, AL 35294-4440
An investigation was conducted on the emissions and efficiency from hydrogen blended compressed natural gas (CNG) in light duty vehicles. The different blends used in this investigation were 0%, 15%, 30%, 50%, and ~100% hydrogen, the remainder being compressed natural gas. The blends were tested using a Ford F-150 truck supplied by Arizona Public Services. Tests on emissions were performed using four different driving condition tests.

Previous investigation by Don Karner and James Frankfort on a similar Ford F-150 using a 30% hydrogen blend showed that there was substantial reduction when compared to gasoline in carbon monoxide (CO), nitrogen oxide (NOx), and carbon dioxide (CO2) emissions while the reduction in hydrocarbon (HC) emissions was minimal.

This investigation was performed using different blends of CNG and hydrogen to evaluate the emissions reducing capabilities associated with the use of the different fuel blends. The results were then tested statistically to confirm or reject the hypotheses on the emission reduction capabilities.

Statistically analysis was performed on the test results to determine whether hydrogen concentration in the HCNG had any effect on the emissions and the fuel efficiency. It was found that emissions from hydrogen blended compressed natural gas were a function of driving condition employed. Emissions were found to be dependent on the concentration of hydrogen in the compressed natural gas fuel blend.

256h Examination of Thermal Performance of Insulation Systems Combining Novel and Conventional Materials for Lh2 Storage

Lixing Gu, Florida Solar Energy Center, University of Central Florida, 1679 Clearlake Road, Cocoa, FL 32922
In general, multilayer insulation (MLI) systems are most commonly used in on-board or small stationary LH2 storage tanks to reduce heat leaks, while perlite insulation systems are most commonly used in large LH2 storage tanks. Both insulation systems require high vacuum levels below 1x10-4 torr. For high vacuum storage tanks, insulation materials are only a small fraction of the total cost. The other manufacturing cost at a high vacuum vessel is much higher than the insulation materials. In order to make cost-effective and robust insulation systems, new insulation systems using combination of conventional insulations and new composite insulation at soft vacuum levels have been proposed. The new systems are expected to have similar thermal performance to conventional insulation system at high vacuum levels and low cost due to requirement of soft vacuum.

The new insulation systems have been tested at boundary temperatures between 77K and 300K in a laboratory. The tested novel materials consist of aerogel composite blanket, layered composite blanket with fiber glass paper and fumed silica dispersion, and layered composite blanket with polyester fabric and fumed silica dispersion. The conventional insulation is MLI. Since test conditions are not the same ones as LH2 storage, it is necessary to examine thermal performance using these novel insulation systems with boundary temperature between 20K and 300K.

The present paper investigates thermal performance for these novel insulation systems at soft vacuum levels through computer simulations. General equations of thermal conductivities for these tested insulation materials were developed first. The equations are a function of temperatures and pressures. Then the models were validated against measured data. As long as the models provide high confidence, the models were applied to LH2 storage conditions and a parametric study was performed. A parametric list consists of following insulation systems at different vacuum levels: • Conventional insulations • Novel insulations • Combination of conventional and novel insulation • Combination of different novel insulations

Simulation results show that there is no benefit using combination of different insulation materials at a certain vacuum level, because there insulations require different vacuum levels to make the best thermal performance, unless a vacuum level may be varied during storage.

NANO-SCALE INVESTIGATION OF MORPHOLOGIES IN POLYMER ELECTROLYTE/PVDF BLEND MEMBRANES

James T. Goldbach, Arkema Inc., King of Prussia, PA
Karren L. More, Oak Ridge National Laboratory, Oak Ridge, TN
Scott R. Gaboury, Arkema Inc., King of Prussia, PA
Michel Foure, Arkema Inc., King of Prussia, PA
David A. Mountz, Arkema Inc., King of Prussia, PA
Amy L. Manheim, US Department of Energy, Washington, DC

Biography

Dr. Goldbach is a Research Scientist in the Corporate Research department at Arkema Inc., where directs the polymer synthesis group working on the development of novel polyelectrolytes and fuel cell membranes. He earned his Ph.D. in Polymer Science and Engineering from the University of Massachusetts – Amherst in 2003 specializing in polymer synthesis and block copolymer thin film morphology characterization.

Dr. More is a Distinguished Research Staff Member at Oak Ridge National Laboratory, where she is currently Acting Group Leader for the Microscopy, Microanalysis, & Microstructures Group in the Materials Science & Technology Division. She earned her Ph.D. in 1992 from North Carolina State University in Materials Science and Engineering.

Dr. Gaboury is a Research Manager for fluoropolymer coatings and films at Arkema Inc., where he supervises the Fuel Cell Membrane team and the general development of PVDF/polyelectrolyte blend membranes. He earned his Ph.D. in Polymer Chemistry from North Dakota State University in 1993.

Dr. Foure is a Director of the Fluorochemicals, Hydrogen Peroxide and Fuel Cell Membranes research at Arkema Inc. He earned his Ph.D. in Organic Chemistry from The University of Wuerzburg in 1975.

Dr. Mountz is a Research Scientist in the Corporate Research department at Arkema Inc., where he directs the membrane formulation group working on the optimization, testing, and development of novel fuel cell membranes. He earned his Ph.D. in Polymer Science and Engineering from the University of Southern Mississippi in 2002 specializing in polyelectrolyte membranes.

Ms. Manheim is a Technology Development Manager at the Department of Energy's Hydrogen, Fuel Cells and Infrastructure Program. She has worked for the Federal Government for almost 20 years and currently manages about $20 million of fuel cell R&D annually. Prior to working for the Government, Ms. Manheim spent over 15 years as an organic chemist conducting *polymer and pharmaceutical research. She earned a BA degree from Northwestern University and an MBA from Virginia Tech.*

Abstract

The polymer electrolyte membrane (PEM) fuel cell is a technology integral to the overall transition from a petroleum-based economy to a renewable resource/hydrogen based economy. At the heart of this essential device is a highly specialized polymer membrane that has been tailored to have high proton conductivity, while maintaining low permeability to the fuel gases and low electrical conductivity. During extended operation in a fuel cell environment, degradation of the membrane can occur **(1)**, contributing to the ultimate failure of the PEM fuel cell. The morphologies of a series of novel polymer electrolyte/poly(vinylidene fluoride) (PVDF) blend membranes have been evaluated using transmission electron microscopy (TEM), both before and after incorporation of the membrane into a membrane electrode assembly (MEA) and following electrochemical aging of the MEA in a fuel cell. In particular, nanometer-scale changes to the polymer structure and changes in spatial distributions of fluorine, sulfur, and other elements were analyzed. The effect of microstructural and microchemical observations has been correlated to a decline in fuel cell performance with time.

Introduction

The current state-of-the-art materials that compose fuel cell membranes fall within the class of perfluorinated sulfonic acid (PFSA) copolymers. All of these copolymers consist of a single copolymer of tetrafluoroethylene (TFE) with a sulfonic acid-functionalized, perfluorinated vinyl ether monomer. The most common of these materials is Nafion®, a material originally designed for, and still mainly used in, the chlor-alkali industry. Figure 1 shows the synthesis of Nafion's® constituent perfluorinated vinyl ether monomer, its copolymerization with TFE, and the subsequent processing into an acidified polymeric material.**(2)** Starting with TFE and sulfur trioxide, there are seven distinct steps which must be carried out to obtain Nafion® in a form useful as a polymer membrane. This multi-step manufacturing process results in a relatively high membrane cost. In addition, the Nafion® synthesis procedure is considered static in that, if any modification of the polymer structure is required, many or all of the synthesis steps may need to be revised. This lack of synthesis versatility results in a limited number of possibilities for modification of the chemical and physical properties of Nafion®.

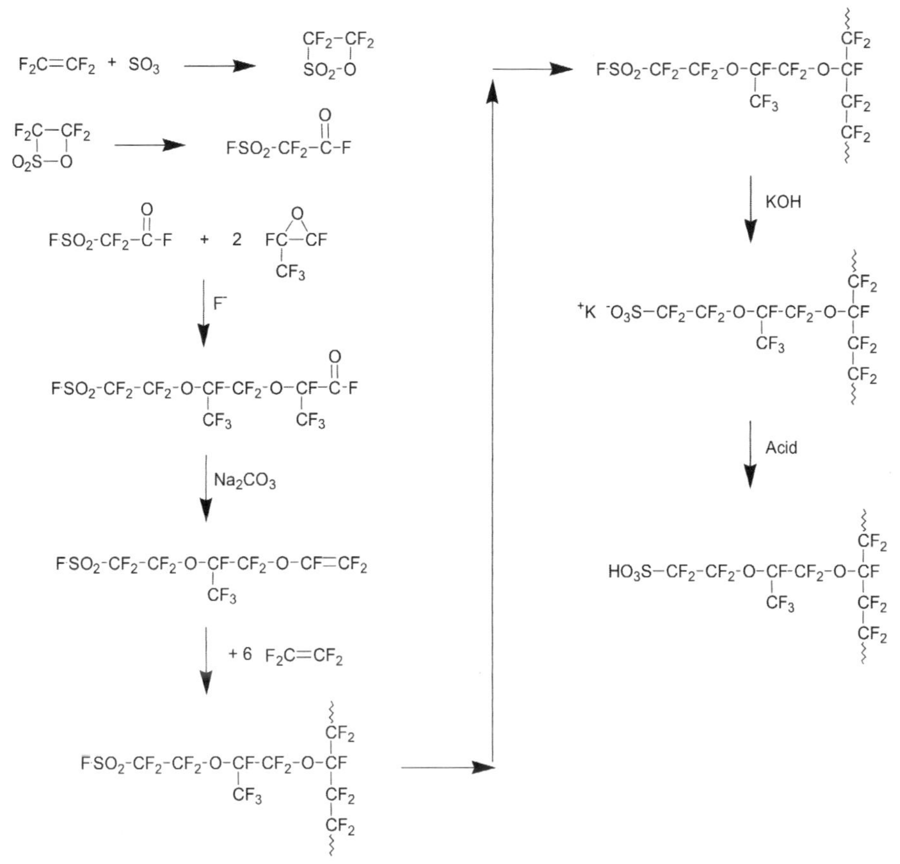

Figure 1. Synthesis of Nafion®.(1)

Arkema's Membrane Fabrication Process

For the past three years, researchers at Arkema Inc., supported by a U.S. Department of Energy grant, have been developing a fuel cell membrane using a novel approach depicted in Figure 2. This approach consists of a matrix (co)polymer and a polyelectrolyte which are combined into a single polymeric material through a proprietary process. For this publication, the matrix copolymer used is Kynar® poly(vinylidene fluoride) (PVDF) for several reasons. First, the use of this type of approach and a commercially-available matrix material such as Kynar® as the primary component of the resulting blend is likely to reduce the cost of the final membrane. In addition, the Kynar® component provides stable mechanical, thermal, chemical, and electrochemical properties for the membrane.**(2)** Unfortunately, Kynar® is not proton-conducting, making it unsuitable to use by itself as a fuel cell membrane.

As outlined in Figure 2, researchers at Arkema Inc. have developed a process whereby matrix copolymers such as Kynar® PVDF can be blended with a proton-conducting polyelectrolyte. The addition of the polyelectrolyte component brings the water absorption and proton conductivity necessary for use of the final blended polymer membrane in a PEM fuel cell. This approach not only has the advantage of utilizing lower-cost materials, but also of being a convergent, blending process, where the two components are synthesized independently and then combined. As opposed to the conventional, static process, this convergent process allows for a nearly unlimited number of component blending ratios and variations of matrix and proton-conducting copolymers.

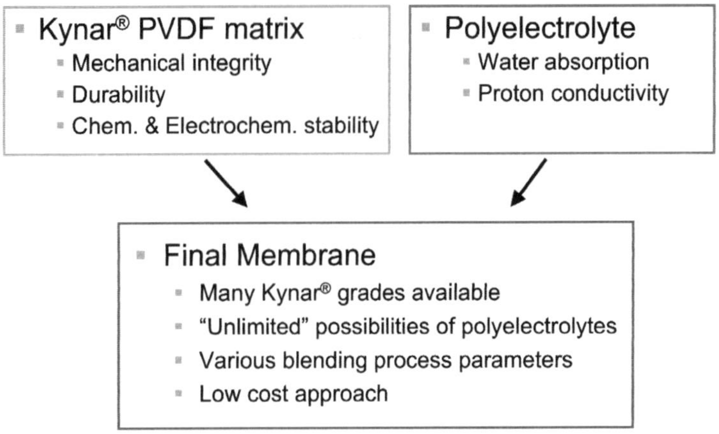

Figure 2. Arkema's convergent approach to fuel cell membranes

The morphology of the final membranes will significantly contribute to the final proton conductivity. Figure 3 shows representative scanning electron microscopy (SEM) images of Arkema Kynar®/polyelectrolyte blend membranes prepared with varying amounts of blending aids. The membrane showing gross separation of the two phases (3a) exhibits weak proton conductivity and poor in-cell performance. However, when the two components are more intimately blended using a proprietary compatibilizing technology, (as in 3b) and no phase separation is discernable, the proton conductivity and in-cell performance are greatly increased. In order to fully understand the correlation of membrane morphology with fuel cell performance, improved imaging and compositional/chemical resolution within the membrane in these intimately blended polymer membranes is required.

Characterization of PFSA Membranes

The properties of Nafion® membranes have been extensively characterized including, but not limited to, electrochemical performance, gas and liquid cross-over measurements, and chemical degradation. **(4-6)** A particularly difficult task has been to accurately characterize the polymer morphology present in any given Nafion® sample. Since Nafion® contains perfluorinated and sulfonated groups (Figure 1), it exhibits both hydrophilic and hydrophobic character. It also changes physical dimensions (swells) with changing relative humidity, which may be accompanied by microstructural changes within the membrane. It has proven difficult thus far to obtain an accurate picture of the structure of the polymer network and/or the exact arrangement of ionic groups. Research has been conducted using small-angle x-ray scattering (SAXS), scanning electron microscopy (SEM) and transmission electron microscopy (TEM) in order to characterize the atomic structure of Nafion® membranes **(5)**.

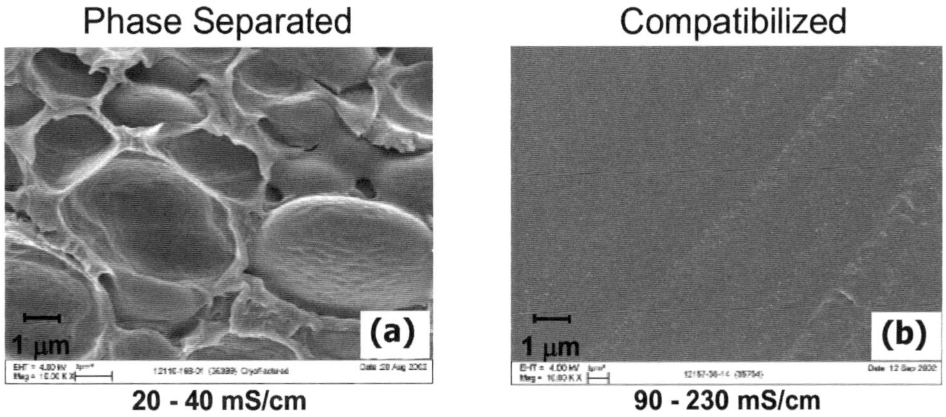

Figure 3. SEM images of Arkema polyelectrolyte/PVDF blend membranes produced (a) without and (b) with Arkema's patented blending process.

Transmission electron microscopy (TEM) of diamond-knife microtomed cross-sections has been used to characterize the nanometer-scale structure and composition of Arkema's M31-generation membranes. For different polymer blends, or phase-separated block copolymers, TEM (coupled with microanalysis techniques such as energy dispersive spectrometry (EDS) and electron energy loss spectroscopy (EELS)) can provide critical information regarding the size, composition, chemical make-up, and crystalline nature of the individual phase-separated regions comprising the membranes. Thus, in addition to high-resolution images, TEM technology can provide high spatial-resolution analytical data to correlate the structural, compositional, and chemical observations for each phase within the different blended polymer membranes, and how each phase changes during electrochemical aging of a fuel cell. TEM can also be used to characterize the presence of, and chemical nature of phase-separated morphologies present at numerous stages of the blending process; information that can then be utilized to further optimize membrane processing.

Figure 4 shows a low-magnification SEM image of a full cross-section of an Arkema M31-generation membrane after a water-wash. Readily visible in the SEM image are alternating light-dark contrasted bands (<1 μm thick) which run perpendicular to the plane (surface) of the membrane. Because these bands are not observable in the as-cast membrane (not shown), it is theorized that they are actually diffusion channels created when the as-cast membranes are water-washed. Compositional analyses (by EDS) reveal that the level of sulfur (S) is slightly higher and the level of fluorine (F) is slightly lower within the light bands, while the levels of S and F vary oppositely within the darker bands. Thus, the bands (pathways) result from regular modulations in S and F content, which strongly suggests that the movement of water through the membrane causes a localization of sulfur-containing polyelectrolyte phase. This evidence implies that highly proton-conductive pathways can be created perpendicular to the plane of the membrane, which could explain the enhanced fuel cell performance observed for intimately blended membranes, versus those showing gross phase separation.

Figure 4. Low magnification image of a cross-sectioned M31 membrane after water wash. Diffusion pathways run perpendicular to the plane of the membrane.

In addition to water diffusion pathways, small (~100 nm), dense nodules (brightly-imaging features in Figure 4) are also observable throughout the M31 membrane through the use of TEM technology. Conventional TEM imaging of the nodules (not pictured) revealed amorphous structures similar to the matrix polymer; however, Figure 5 clearly shows these small nodules as having a different composition (evidenced by contrast differences in the SEM images). The reason for this difference lies in the fact that scanning TEM (STEM) imaging and analysis with a small electron probe (~2 nm diameter) was used to characterize the fine structure of the M31 membrane in Figure 5, and was combined with elemental mapping using EELS to elucidate fine-scale compositional differences among the phases. An annular dark-field (ADF) STEM image is shown in Figure 5 along with elemental maps of the different membrane constituents. The nodules were identified as N-enriched, and S-, O-, and F-depleted. The carbon map shows that a mass-thickness effect (density difference due to localized changes in C bonding, i.e. the chemical nature of C) was observed, and can be associated not only with the N-rich nodules, but also with thin, dense, linear features, which run parallel to the membrane surface and also display brightly in the STEM image.

The TEM, STEM and EELS analyses have proven invaluable for the characterization of the compositional and chemical nature of various phases/regions present in Arkema M31-generation fuel cell membranes in both the as-cast condition and after water washing. However, the fabrication of a useable PEM fuel cell membrane electrode assembly (MEA) requires the application of porous catalyst layers to the membrane surfaces and the pressing of the membrane and gas diffusion layers (GDLs) to fabricate a finished product. These additional processing steps, involving hot-pressing and elevated temperatures, can have a drastic and sometimes detrimental effect on the resulting microstructures of fuel cell membranes.

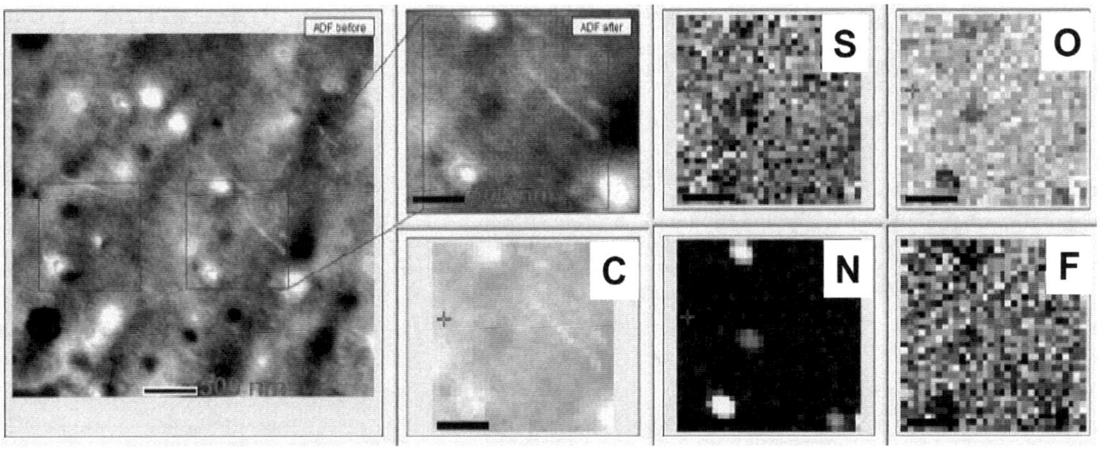

Figure 5. High magnification STEM image and associated EELS elemental maps of S, O, C, N, and F. High density nodules show increased N and decreased O and F contents.

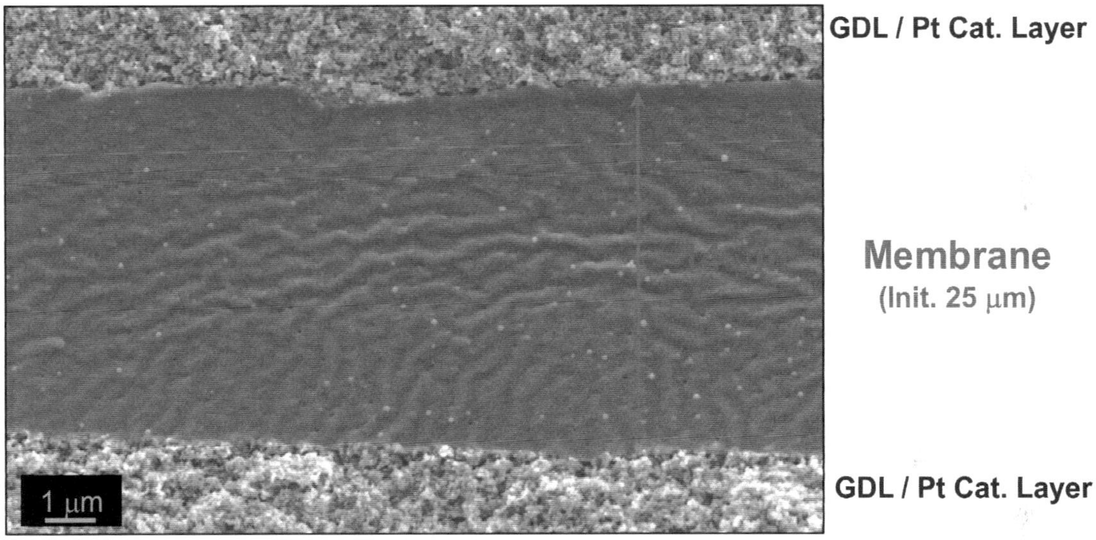

Figure 6. Low magnification SEM image of M31 as-processed MEA. Membrane thickness decreased from 25μm to 14μm. The water 'diffusion pathways' have been deformed.

TEM analysis can also be used to evaluate diamond-knife microtomed cross sections of MEAs in an effort to further analyze the effects of this final step in the membrane-creation process. Figure 6 shows a cross-section SEM image of an MEA with an Arkema M31 membrane. As in previous SEM images, the water diffusion pathways are clearly visible as well as the nitrogen-rich nodules. However, it can be observed that the overall membrane thickness has been reduced from an original thickness of 25 μm to 14 μm. In addition, the diffusion pathways no longer run directly perpendicular to the plane of the membrane, but are instead bent toward the center. This visual evidence suggests that the pressing conditions used in MEA preparation were too harsh and severely compressed the membrane. Using this data, the new MEA processing conditions were modified to keep the membrane at its original 25 μm thickness.

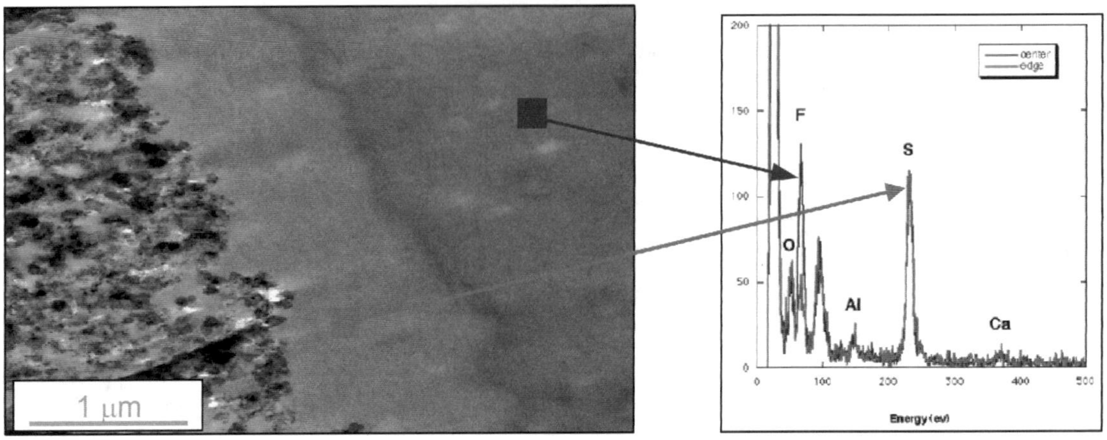

Figure 7. High-magnification TEM image of the electrode-membrane interface. A ~1μm sulfur-enriched, F-depleted layer is evident within the membrane adjacent to the electrode. EDS data are normalized with respect to S-K peak.

High-magnification STEM imaging was also used to analyze an MEA. Figure 7 shows the membrane-electrode interface region in an as-processed MEA. An interface layer within the Arkema M31 membrane was observed in the region adjacent to the Pt-electrode. EDS analysis of this region shows that there is an increase in the S and a decrease in the F content, as compared to the bulk composition of the M31 membrane. These results suggest a localization of the polyelectrolyte phase at the membrane surface adjacent to the Pt electrode in the MEA. Two main conclusions may be drawn from this observation: one, that the polyelectrolyte phase has some local mobility during the MEA preparation step, and two, that the Kynar® PVDF does not prefer to lie at the electrode interface. The information obtained using STEM imaging will be used for the development of particular polyelectrolyte blend layers, which are specifically designed to enhance binding properties at the electrode-membrane interface.

Similar imaging techniques were applied to an Arkema M31 MEA that had been aged for 2100 hours in a fuel cell. In addition to the S-enriched GDL interface region observed in BOL MEA samples, additional, segregated S-enriched regions were observed throughout the membrane (Figure 8). TEM images of the EOL MEA show that the modulated structure of the membrane (alternating high-S, low-F bands) and the dispersed N-rich nodules did not change during testing, however additional S-rich phase segregation occurred, which was also associated with the existing modulated structure. This indicates phase instability associated with the polyelectrolyte phase. Work is ongoing to modify, test and characterize new, more stable generations of PVDF/polyelectrolyte blend membranes. TEM results for these materials will be similarly correlated to in-cell performance and used in further optimization iterations.

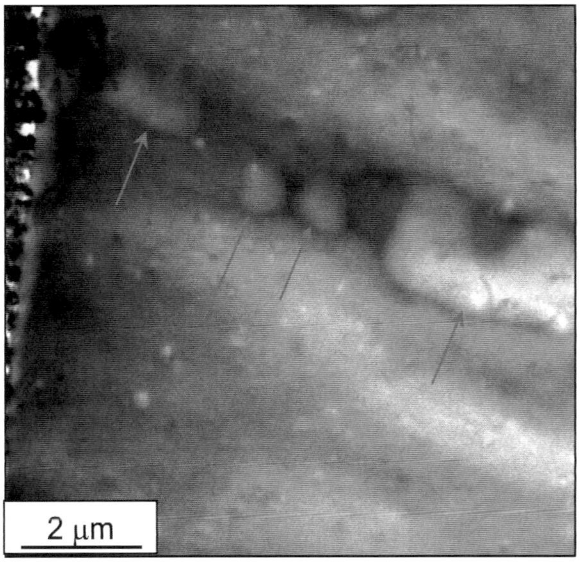

Figure 8. TEM image of EOL M31 MEA. Red arrows identify S-rich phases not present in BOL samples.

Conclusions

Researchers at Arkema Inc. have developed a robust process by which nearly any polyelectrolyte can be blended with PVDF or other matrix (co)polymer to form a fuel cell membrane. This process has the advantage of utilizing lower-priced materials to achieve unlimited tenability in membrane properties by adjusting blending ratios or the natures of the matrix copolymer and polyelectrolyte. TEM-ADF has been applied to these membranes in as-cast as well as processed forms to elucidate not only morphological characteristics about the polymer blend, but also to gain knowledge of the chemical natures of each feature in the TEM image. This knowledge is fed back into the process optimization to reduce impurity concentration and optimize *ex-situ* proton conductivity. The TEM-ADF technique has also been successfully implemented to the characterization of MEAs, and the structural and chemical information gained has been utilized in the optimization of MEA preparation procedures as well as the characterization of GDL-membrane interfacial regions. Continued use of this technique will allow for accelerated development and optimization of the Arkema PVDF/polyelectrolyte blend fuel cell membrane.

References

1. Curtin, D. E., Lousenberg, R. D., Henry, T. J., Tangeman, P. C., Tisack, M. E. (2004), "Advanced Materials for Improved PEMFC Performance and Life," *Journal of Power Sources*, pp. 41-48.

2. Connolly, D. J., Gresham, W. F. (1966), US Patent 3282875, "Fluorocarbon Vinyl Ether Copolymers", E. I. du Pont de Nemours & Co. assignee.

3. Arkema Inc. Kynar® technical bulletin: www.arkemagroup.com

4. Yu, J., Matsura, T., Yoshikawa, Y., Islam, M. N., Hori, M., (2005), "*In Situ* Analysis of Performance Degradation of a PEMFC under Nonsaturated Humidification," *Electrochem. Solid-State Lett.*, 8(3), ppA156-A158.

5. Ise, M., Diat, O., Gebel, G., Kreuer, K.D., Maier, J., (1999), "Comparative SAXS Study of Nafion and Sulfonated PEEKK Membranes," *Hamburger Synchrotronstrahlungslabor Annual Report-1999*. (http://www-hasylab.desy.de/science/annual_reports/1999_report/index.html)

6. Zhou, C., Schiraldi, D.A., Zawodzinski, T.A., (2005), "Studies of Model Compounds and Nafion® Under Mimic Fuel Cell Conditions: Chemical Durability," *Polymer Preprints*, 46(1), pp. 327-328.

Membrane Desulfurization of Logistic Fuels for Fuel Cell Auxiliary Power Units

Kandaswamy DuraiSwamy, Richard Root Woods, and Diane Aagesen.
Intelligent Energy Inc. 2955 Redondo Ave. Long Beach, California 90806

In recent years, industry experts have begun to consider fuel cells as viable alternatives for auxiliary power generation in commercial aircraft, civilian and military vehicles, along with the largest ships used in defense and transportation. Some of the main problems with conventional auxiliary power units (APU) include their poor efficiency and added contribution to environmental degradation. Modern, high efficiency fuel cells proposed for auxiliary power generation are typically constructed from components that are quite sulfur intolerant.

To enable new fuel cell APU technology, it becomes imperative to reduce sulfur level concentrations in jet fuels and logistic fuels for subsequent on-board, on demand power production. Primary logistic fuels and jet fuels usually contain between 300ppm to 5,000ppm heavy-sulfur compound levels that must be processed such that a clean fuel stream contains no more than 1ppm sulfur. Today, most fuel desulfurization technology under development employs sorbent beds that undergo cyclic sorption and regeneration steps requiring thermal cycling that inevitably wastes fuel in the regeneration combustion step.

Intelligent Energy (IE) Inc. has developed a novel and compact membrane desulfurization technique that does not waste fuel and has an extremely low parasitic power demand. Pervaporation as the process is known, separates a primary fuel mixture into essentially a sulfur free stream for a fuel cell sub-system, and a sulfur enriched fuel that is returned to the main fuel supply tank. The Intelligent Energy multi-stage membrane approach to desulfurization is more compact and less energy intensive than adsorbent and distillation type sulfur reduction methods used by competitors.

The following paper presents design aspects of a two-stage scaleable membrane desulfurization process along with experimental data obtained from commercial jet fuel (Jet-A) processing trials. It has been determined that the success of the IE approach depends largely on producing a light, gasoline-range sulfur intermediate stream that serves as the feedstock to a second, more highly sulfur-selective stage. A comparative survey of various desulfurization methods is examined and the paper concludes with data that validate the proof-of-concept of the aforementioned approach.

Slide 1

On-Board Membrane Desulfurization for Fuel Cell Auxiliary Power Units

Presented by:
Intelligent Energy Inc.
Kandaswamy DuraiSwamy
Richard Root Woods
Diane Aagesen

AIChE National Spring Meeting
April 27, 2006
Orlando, Florida

Slide 2

The Various Methods of Sulfur Reduction in Hydrocarbon Fuels

- **Distillation**
- **Nano-filtration**
- **Chemical Adsorption**
- **Hydrotreatment**
- **Pervaporation**

Slide 3

What is Pervaporation?

- Pervaporation is a process that separates mixtures of liquids by partial vaporization through a non-porous membrane
- The membrane acts as a sulfur selective barrier between two phases, the liquid phase feed and the vapor phase permeate
- The fractionation/separation is achieved mainly due differences in polarity between compounds rather than differences in relative volatility

Slide 4

Applications of On-Board Desulfurization

- **Fuel Cell Auxiliary Power Units for....**
  - **Aircrafts** –Military and Civilian (75-100kW)

    Boeing is a key sponsor to Intelligent Energy Desulfurization Project
  - **Tanks**-Military (75-100kW)
  - **Trucks**-Commercial (5-25 kW)
  - **RVs**-Commercial (5-25 kW)
  - **Ships** –Navy and Merchant (100-300kW)

Slide 5

Some Benefits of Fuel Cell Auxiliary Power Units

	Annual Fuel Savings	Annual Emissions Reduction
Commercial Aircrafts 	½% in-flight fuel 876,000,000 gallons at airports	9.3 million tons CO_2 at airports >99% NOx and particulate matter reduction
Commercial Class 8 Trucks	419,000,000 gallons	4.64 million tons CO_2

Slide 6

Intelligent Energy's Novel 2 Stage Desulfurization Process (10kW System)

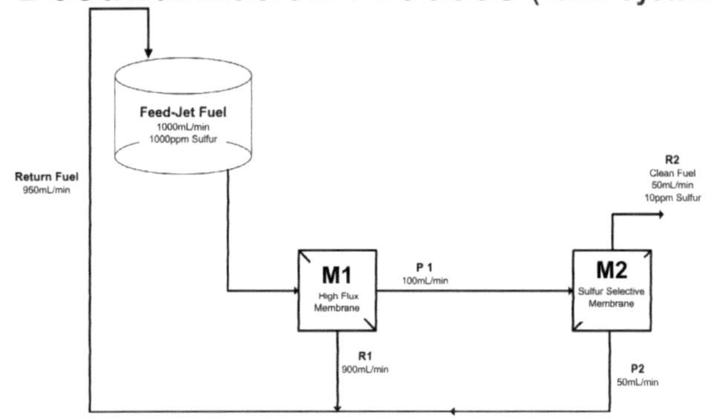

Slide 7

Distillation vs. Pervaporation

- •Pervaporation is more energy efficient than distillation
- •Pervaporation more effectively produces a light fuel
- •Pervaporation is more compact and not orientation dependent

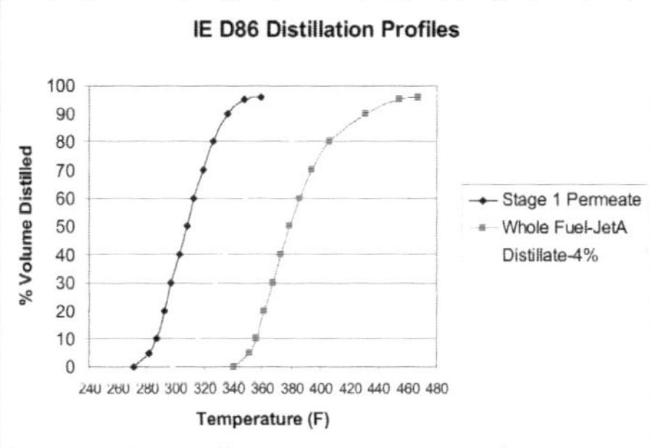

Slide 8

Multi-Stage Desulfurization System Performance

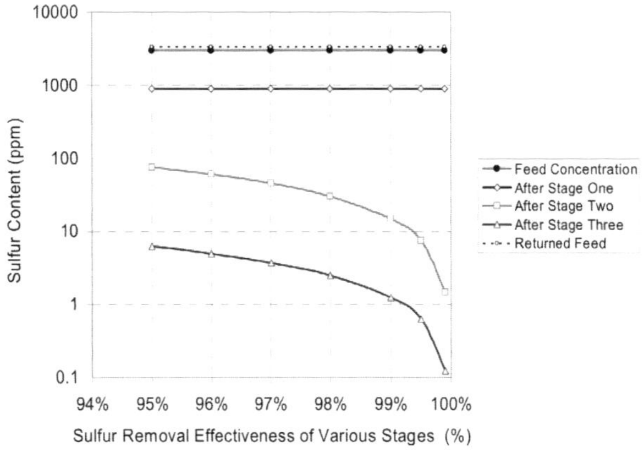

Slide 9

Packaged Membrane Desulfurization Hardware

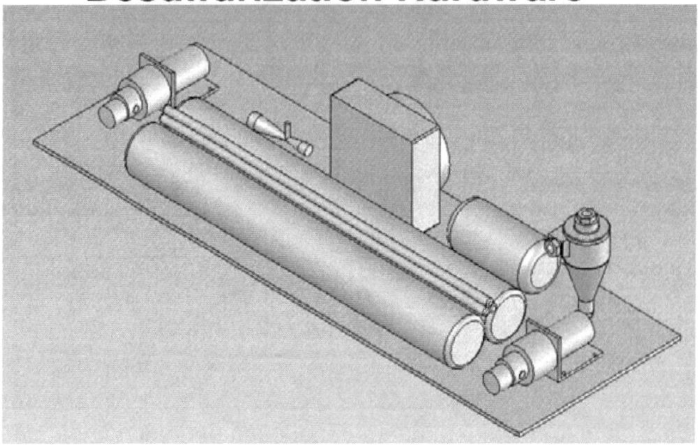

Slide 10

Conclusions

- Sulfur content in jet fuel is a critical limiting factor for fuel cell based auxiliary power units
- Multi-stage on-board desulfurization of fuel for auxiliary power units is technically feasible.
- 1st and 2nd stages using pervaporation membranes with an optional polishing filter (3rd stage) can bring the sulfur level in the clean fuel down to approximately 1 ppm.

268c Hydrocarbon Membranes Composed of Poly(Vinyl Alcohol) Derivatives for the Application of Fuel Cell

Sang Yong Nam, Department of Polymer Science and Engineering, Gyeongsang National University, 900 Gazwa-dong, Jinju, South Korea, Ji Won Rhim, Department of Chemical Engineering, Hannam University, Daejeon, South Korea, Hoon Huh, Korea Institute of Industrial Technology, 35-3 Hongchon-Ri, Ipchang-Myun,, Chungnam, Chonan-Si 330-825, Korea, Chonan-Si, South Korea, and Seong Yong Ha, Airrane, 4F, Aekyung Institute, 217-2 Shinsung-dong, Daejeon, South Korea.

A polymer electrolyte membranes fuel cell are one of the most attractive power sources for a wide range of application from vehicles to portable utilities due to the stable operation at a rarely low temperature, the high energy generation yield and energy density, the simplicity of system. PVA is a water-soluble polyhydroxy polymer that is used in practical applications because of its easy preparation, excellent chemical resistance, and physical properties. In particular, PVA membrane has been used in ethanol dehydration to break the ethanol-water azeotrope because it selectively passes water molecule over ethanol or methanol. Although PVA itself dose not have the fixed charges, several organic groups like hydroxyl, amine, carboxylate, sulfonate, and quaternary ammonium can be used to impart hydrophilicity and/or ionic group. As reported earlier, very hydrophilic membranes for pervaporation were evaluated as polymer electrolyte membrane (PEM) in fuel cells application. In this study, crosslinked poly(vinyl alcohol) (PVA) membranes were prepared at various crosslinking methods containing sulfuric acid groups.

268d Development of Dense Cermet Membranes for Hydrogen Separation

U. (Balu) Balachandran, Tae H. Lee, Ling Chen, Sun-Ju Song, and Stephen E. Dorris. Energy Technology, Argonne National Laboratory, 9700 S. Cass Avenue, Argonne, IL 60439

Argonne National Laboratory (ANL) is developing dense ceramic membranes for separating hydrogen from mixed gases, particularly product streams generated during coal gasification and/or methane reforming. Hydrogen separation with these membranes is nongalvanic, i.e., it does not use electrodes or an external power supply to drive the separation, and hydrogen separated from the feed stream is of high purity, so post separation purification steps are unnecessary. Dense cermet (i.e., ceramic-metal composite) membranes have been developed to separate hydrogen from gas mixtures at high temperature and pressure. The hydrogen permeation rate, or flux, measured at atmospheric pressure in the temperature range of 500-900°C, varied linearly with the inverse of membrane thickness, and reached ≈19 cm^3[STP]/min-cm^{2}at 900°C for an ≈20-μm-thick membrane when 100% $H_2$was the feed gas. The effects of membrane thickness and hydrogen partial pressure on flux indicated that bulk diffusion of hydrogen is the rate-limiting step. A thin film of the cermet membrane on a porous support structure gave a flux of ≈33 cm^3[STP]/min-cm^2 at 900°C when 100% H_2 was the feed gas. The hydrogen flux also varied with the microstructure of the cermet. When some of the membranes were tested in a gas mixture that contained high concentrations of CH_4, CO and CO_2 for times that approached ≈500 h, no performance degradation was observed. Hydrogen flux measurements in H_2S-containing atmospheres show that the cermet membranes are stable for up to 1200 h in gases that contain 400 ppm H_2S. These results suggest that the membranes are chemically stable and may be suitable for long-term operation. The present status of membrane development at ANL will be presented in this talk. Work supported by U.S. Department of Energy, Office of Fossil Energy, National Energy Technology Laboratory's Hydrogen and Gasification Technologies Program, under Contract W-31-109-Eng-38.

268e Performance of Pd-Alloy Membranes for Hydrogen Separation from Mixed Feed Streams Containing 1000 Ppm H2s

Osemwengie U. Iyoha, National Energy Technology Laboratory, Department of Chemical and Petroleum Engineering, University of Pittsburgh, 1249 Benedum Hall-University of Pittsburgh, Pittsburgh, PA 15261, Bryan M. Morreale, Parsons Project Services Inc., P.O. Box 10940, Pittsburgh, PA 15236, Richard P. Killmeyer, US DOE National Energy Technology Laboratory, P.O. Box 10940, Pittsburgh, PA 15236, Robert M. Enick, University of Pittsburgh, 1249 Benedum Hall, Pittsburgh, PA 15261, and Bret H. Howard, U.S. Department of Energy National Energy Technology Laboratory, P.O. Box 10940, Pittsburgh, PA 15236.

Introduction

There is growing interest in the integration of membrane technology into commercial processes such as coal gasification for simultaneous hydrogen production and separation. Consequently, membrane materials will be exposed to varying concentrations of impurities in the feed gas stream. H_2S is a major concern since it is known to adversely affect the performance of metal-based membranes with respect to hydrogen transport and membrane stability.

Pd-Cu alloys have gained widespread interest and are regarded as promising candidates for membrane reactor applications as a result of their catalytic activity towards the water-gas shift reaction (Bustamante et al., 2005), in addition to potential H_2S tolerance at certain operating conditions and H_2S concentrations (Morreale et al., 2004). As a result, NETL is conducting extensive tests of these materials as part of a program to develop H_2S tolerant, post-gasifier water-gas shift membrane reactors to enhance hydrogen production via coal gasification (Howard et al., 2004).

Results and Discussion

Effect of H_2S

The performance of various Pd-Cu alloy membranes as well as pure Pd membranes have been studied over a range of temperatures in the presence of 1000 ppm H_2S in H_2. Preliminary analysis shows dramatic losses of H_2 flux for 100 µm Pd membranes at moderate temperatures in addition to the growth of a stable sulfide layer on the membrane surface. Figure 1 depicts the decrease in H_2 transport of a pure Pd membrane on exposure to the H_2-H_2S feed stream at 623K and 0.6 MPa. SEM elemental mapping of the membrane cross section revealed a Pd-S layer approximately 18 µm-thick on the membrane surface after a 120 hour exposure. XRD analysis determined the sulfide layer to be Pd_4S. Similar performance decreases were observed for Pd-Cu alloy membranes at moderate temperatures. By contrast, at higher temperatures, preliminary results for some Pd-Cu alloys suggest minimal change in membrane performance.

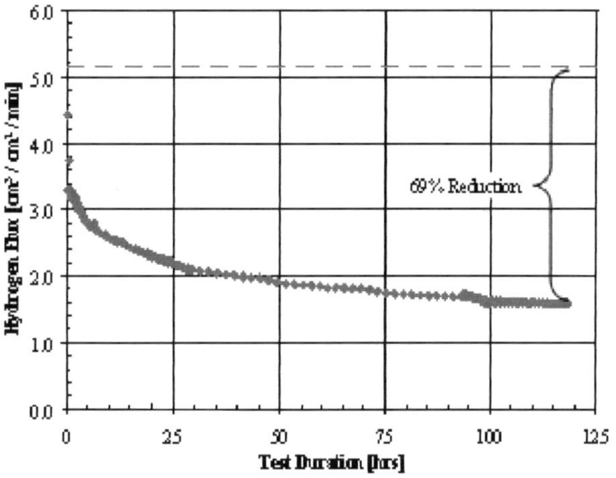

Figure 1. Flux results for a 100 mm Pd membrane exposed to 1000 ppm H_2S–H_2 at 623K as a function of exposure time.

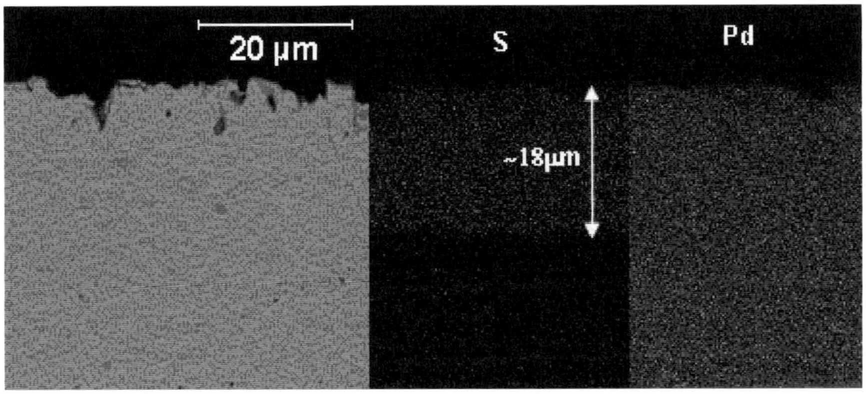

Figure 2. Pd and S elemental maps of the cross section of a 100 µm Pd membrane exposed to 1000 ppm H_2S-H_2 at 623K for 120 hrs.

Effect of CO, CO_2 and H_2O

Experiments to investigate the influence of high CO, CO_2 and H_2O concentrations on 80wt%Pd-Cu alloy membranes were conducted using 1 mm thick membrane samples. Although these samples were too thick to accurately demonstrate the influence of these gases on the H_2 transport properties of the membrane, the experiments revealed valuable information regarding the influence of these components on membrane surface morphology. For example, SEM analysis of the membrane surface revealed significant surface modification after exposure to a 50%H_2-CO feed at 1038K and 1.11 MPa (Figure 3). Similar catalytic surface etching was observed in the H_2-H_2O environment at temperatures between 908 and 1038K. Figure 4 shows the image of an 80wt%Pd-Cu membrane after exposure to the H_2-H_2O environment at 908K and 1.55 MPa.

Figure 3. SEM micrograph of the feed-side surface of the 80wt%Pd-Cu membrane after exposure to a 50%H_2-CO feed at 1038K and 1.11 MPa. The membrane was exposed to the feed stream for 24 hrs. CO_2, CH_4 and H_2O were also present as a result of side reactions.

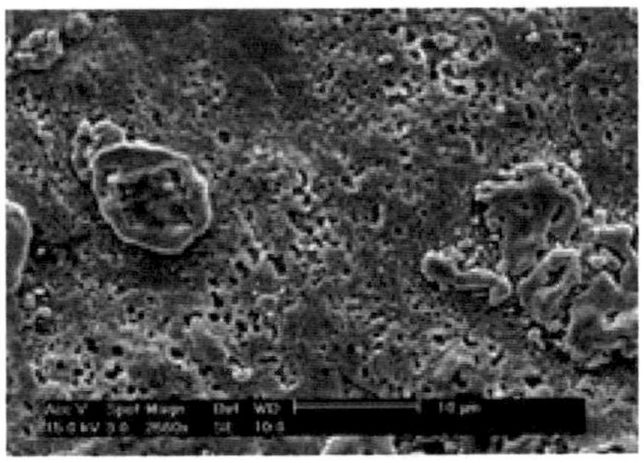

Figure 4. SEM micrograph of 80wt%Pd-Cu membrane after exposure to a 50%H_2-H_2O feed stream at 908K and total unit pressure of 1.55 MPa for 24hrs.

In contrast to the 80wt%Pd-Cu membranes exposed to binary mixtures of H_2 with H_2O, CO or CO_2 feed streams at temperatures between 873 and 1038K which appeared significantly roughened, the membranes exposed to similar feed stream at the higher temperature of 1173K remained relatively smooth (Figure 5).

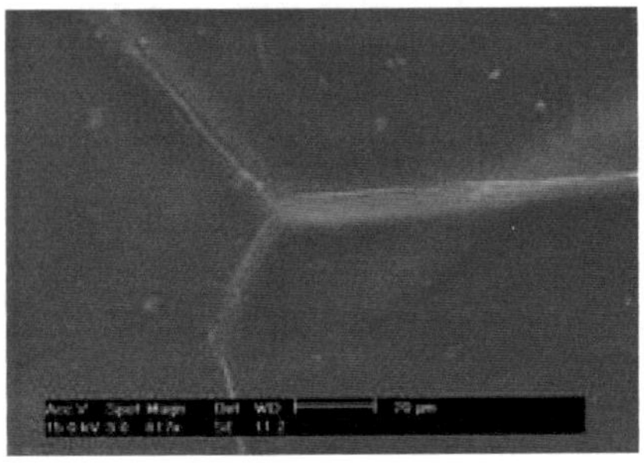

Figure 5. SEM micrograph of the feed-side surface of the 80wt%Pd-Cu membrane after exposure to a 50%H_2-CO feed at 1173K at 1.11 MPa. The membrane was exposed to the feed stream for 24hrs.

Conclusion

Preliminary data for 100 μm thick membranes suggest that in addition to markedly reducing H_2 flux, H_2S impurity levels of 1000 ppm result in corrosion of Pd and some Pd-Cu alloy membranes at low to moderate temperatures. Also, exposing the 80wt%Pd-Cu membranes to CO, CO_2 and H_2O at temperatures between 873 and 1173K resulted in dramatic surface modification. However, the membranes showed minimal adverse surface modification in the presence of the binary mixed feed streams at 1173K. Together, these preliminary results suggest that at the high temperature conditions of post-gasifier water-gas shift membrane reactors, some Pd-Cu compositions may tolerate similar H_2S impurity levels. Future experiments are planned to test this hypothesis.

References

F. Bustamante, R. Enick, R. Killmeyer, B. Howard, K. Rothenberger, A. Cugini, B. Morreale, M. Ciocco. Uncatalyzed and Wall-catalyzed forward water-gas shift reaction kinetics, AIChE J. 51 (2005) 1440

B. Howard, R. Killmeyer, K. Rothenberger, A. Cugini, B. Morreale, R. Enick, F. Bustamante. Hydrogen permeance of palladium-copper alloy membranes over a wide range of temperatures and pressures, Journal of Membrane Science 241 (2004) 207

B. Morreale, M Ciocco, B. Howard, R. Killmeyer, A. Cugini, R. Enick. Effect of hydrogen-sulfide on the hydrogen permeance of Palladium-copper alloys at elevated temperatures, Journal of Membrane Science 241 (2004) 219

ABSTRACT

Hydrogen Purification with Palladium Alloy Membranes

James R. Brenner and Gaurav Bhagat
Florida Tech Chemical Engineering

As the economy migrates from fossil fuels to H_2, membrane purification of hydrogen is becoming increasingly important. Critical to this H_2 purification is the rejection of impurities such as CO, H_2S, and SO_2 to produce an ultra high purity grade (>99.999%) H_2. Pd and Pd-alloy membranes are especially suited for both low and high temperature hydrogen separation applications. Alloying Pd particularly with Ag not only suppresses hydrogen embrittlement, but also increases the permeability of the alloy membrane. The main objective of this work was to develop a membrane via a simultaneous electroless plating of Pd and Ag. Considering the heavy cost of a pure Pd metal membrane, a thin layer of Pd-Ag alloy was developed on a 0.5 μm media grade porous stainless steel to act as a H_2 purifier. Effects of the electroless bath temperature, composition, annealing time and temperature were studied while developing the Pd coating. The topography and the homogeneity of the developed membrane were analyzed via SEM-EDS and AFM. EDS of Pd-Ag layer coated on a porous stainless steel tube showed a 79Pd-21Ag composition. The tube was annealed at 600°C for over 15 hours in an inert gas (N_2) only once after required thickness of deposited metal was obtained. Fick's 2nd law was used to simulate the unsteady-state interdiffusion between Pd and Ag layers during heat treatment. Once the desired Pd-Ag alloy composition was seen, and impermeability of the developed membrane to non-H_2 gases was confirmed, the membrane was fabricated into a shell and tube type purifier module. Very low N_2 fluxes were seen, ranging from 0.010 - 0.020 cm^3/cm^2-min at temperatures varying from 300°C - 400°C. Although the exact source of this modest leak could not be pinpointed, the presence of pinholes or grain boundary diffusion could not be ruled out. H_2 permeance varying from $1.9 \times 10^{-2} - 3.1 \times 10^{-2}$ cm^3/cm^2-min-$Pa^{0.5}$ was recorded between temperatures from 350°C - 550°C. Sievert's Law was seen to hold true at all temperatures, with regression coefficients between 0.95 – 0.99. Long term stability (>15 hrs) and reproducibility of the H_2 permeance was confirmed. Finally, the activation energy (E_d) for H_2 permeance in the Pd-Ag layer was calculated from the permeance data to be 9 ± 1 kJ/mol, which was found to be in close proximity with the values found in the literature.

INTRODUCTION

The effect of synthesis method on long-term stability and selectivity of Pd and Pd alloy membranes on a variety of supports [1-11] has been summarized by Shu [12] and Rothenberger [13-14]. A summary of the H_2 fluxes is in Table 1.

Among the variety of techniques useful for deposition of alloys, electroless plating has become the method of choice due to its uniform deposition on complex shapes and large substrate areas, the hardness and adhesion properties of the deposited film, and its use of relatively simple and inexpensive equipment [15-17]. Electroless deposition is based on the controlled, autocatalytic reduction of a dissolved metallic salt by reducing agents at a substrate interface. Although conceptually straightforward, support quality, surface

Table 1: Pd Membrane Performance

Reference	Year	Pd (μm)	T (K)	$P_{H_2,ret}$ (kPa)	k ($mol/m^2/s/Pa^{0.5}$)	N_{H_2} ($mol/m^2/s$)
Yeung [2]	1999	1.6	623	256	8.8E-05	2.8E-02
Shu [15]	1996	2	573	105-120	1.0E-03	3.1E-01
Yan [18]	1994	2	773	100-200	5.0E-06	5.0E-01
Uemiya [19]	1991	20	673	150-394	1.1E-05	7.3E-02
Ma [20]	2003	21.3	623	101-202	3.3E-04	1.0E-01
Lin [21]	1998	22	623	202-1515	1.1E-04	3.7E-02
Mardilovich [22]	1998	24	623	151-404	3.0E-02	9.8E-02
Tosti [23]	2000	50	623	200	2.1E-04	6.8E-02
Jarosch [24]	2001	156	1023	425-603	2.6E-05	8.4E-03

Notes: $P_{H_2,permeate}$ = 101.3 kPa, $P_{H_2,sweep}$ = 0 kPa.

activation methods, electroless plating procedures and bath chemistry influence membrane selectivity, permeability and Pd/Pd-alloy film stability. Electrolessly plated Pd-Ag membranes can be prepared either by a *coating and diffusion* of separate Pd and Ag layers, or by *simultaneous co-plating* of Pd and Ag from the same plating bath, followed by an annealing process. After forming a series of alternating layers of Pd and Ag via the coating process, the diffusion necessary to achieve a homogeneous composition involves annealing in H_2 at very high temperatures (800-900°C) for many hours [25-27]. Lower annealing temperatures results in a compositional gradient across the membrane [26], which can lead to poorer mechanical stability and lower permeation rates. Moreover, the high temperature necessary for such diffusion would negatively impact the porosity of the porous stainless steel (PSS) support. Since the deposited Pd and Ag grains are in very close proximity, co-plating allows for the formation of homogeneous alloys at much lower temperatures. The complex kinetics involved in the co-plating bath solution of Pd-Ag and the sensitivity of the Pd-Ag film composition on the composition and temperature of the plating bath solution has been the primary focus of other investigators [21-22, 25-28]. At fixed plating temperatures of 328 K or 333 K and $[N_2H_4]$ = 10 mM, when the Ag concentration was ≥ 10 at. % in the plating solution, the crystallite diameter is governed by Ag deposition kinetics, and Pd agglomerates around the Ag [28]. Heating at 500°C for 8 hours resulted in a bimetallic Pd-Ag alloy [28]. *In the experimental procedure for Pd-Ag coating by simultaneous electroless plating, the following parameters were fixed: [Metal] = 10 mM, $[N_2H_4]$ = 10 mM, $[NH_4OH]$ = 3 M, pH = 10.4, and T= 60°C* [28].

EXPERIMENTAL

The procedure below for deposition of Pd and Ag on porous stainless steel was based on those described in Refs. [25] and [28].

1. 0.5 μm grade ¼" OD porous 316L stainless steel tube welded to porous stainless steel (PSS) tubes on both ends were purchased from Mott Corporation.
2. Any surface contaminants were removed by cleaning the tube in an ultrasonic bath with 1 M KOH solution at 60°C for half an hour. This cleaning procedure was followed by sequentially rinsing the tube in DI water and isopropanol.
3. The tube was treated for 5 min. with HCl, then rinsed with distilled H_2O.

4. After loading the SS tube into the tube furnace and wrapping it with quartz wool, the reactor was leak tested with dry air, then purged with 30 sccm dry air at room temperature for 5 min., before ramping up to 600°C at 3°C/min.. After a four-hour calcination treatment at 600°C, the samples were cooled in flowing dry air to 25°C.
5. The oxidized tube was activated by immersing the tube in baths of acidic $SnCl_2$ (1 g/L) and acidic $Pd(NH_3)_4Cl_2{\cdot}H_2O$ (0.15 g/L). The immersion treatments were repeated 4-10 times (depending upon the amount of activation required) for 5 minutes each, and then the activated tube was dried for 2 hours at 120°C. After immersion in $SnCl_2$, each SS tube was rinsed in DI water. Then, after immersion in $Pd(NH_3)_4Cl_2{\cdot}H_2O$, each SS tube was rinsed with 0.01 M HCl and DI H_2O.
6. The activated tube was immersed in the following electroless bath compositions (Table 2) at room temperature, with constant amounts of $Na_2EDTA{\cdot}2H_2O$ (40.1 g/L), 29% NH_4OH (200 mL/L), and 1 M N_2H_4 (10 mL/L), and a constant pH = 10.4. After depletion of each plating solution, a new solution was provided, and the procedure was repeated after rinsing the membrane.

Table 2: Electroless Bath Composition

	Bath 1(only Pd)	Bath 2(only Ag)	Bath 3 (75Pd -25Ag)
$Pd(NH_3)_4Cl_2{\cdot}H_2O$, g/L	2.63	-	2.11
$AgNO_3$, g/L	-	1.69	0.34

7. Then the plating solution and tube were placed in a water bath at 60°C. Each electroless plating bath solution was replaced with fresh solution after every 90 minutes, and each sample was plated 12 times. Between changing of plating bath solutions, each plated tube was rinsed in DI water for 2-5 minutes at 60°C. After 5 minutes of Pd predeposition, the Ag was added for the codeposition [25, 28].
8. To obtain the final Pd-Ag alloy, the palladium-silver/PSS membrane was annealed in flowing hydrogen at 400°C for 4 hours and 500°C for 4 hours.

Samples were then characterized using scanning electron microscopy (SEM), energy dispersive spectroscopy (EDS), and atomic force microscopy (AFM).

The purifier was a 316 SS 1-1/2 Schedule 40 6.08" long pipe with 0.31" thick end plugs on both sides. H_2 entered via the 0.5 µm grade 1/4" OD porous 316 stainless steel (PSS) tubes welded to NPSS on both ends. Two 0.260" ID wells for 1/4" OD, 5" long, 305 Watt electric cartridge heaters were used for heating. A thermowell with a 0.192" ID housed a 3/16" OD 3-point K-type thermocouple that sensed temperature just past the entrance, in the middle, and just prior to the purifier exit. The entire purifier was surrounded with 2" thick quartz wool insulation. Permeation of H_2 and N_2 was measured at temperature ranges 573 - 823 K. Flow rates and pressures were measured using LabView 7.0.

RESULTS AND DISCUSSION

Structural and Chemical Properties

The first Pd-Ag/PSS sample resulted in a non-uniform sample. More critically, there were cracks in the Pd-Ag layer, and the color looked like that of pure Ag. The non-uniformity and cracking in the Pd-Ag layer could be explained by improper control of the electroless bath solution temperature and pH [25, 28]. The heterogeneities on the ends were due to handling without gloves after each plating. For this reason, all handling was done on the nonporous ends of the PSS, the bath pH was controlled at 10.4, the bath solution was set to 60 ± 1°C, and the supports were anchored to a rotating shaft during deposition to improve uniformity. To verify N_2 impermeability, each sample was leak checked with N_2 at 25°C before welding to a purifier. SEM (Fig. 1) and AFM analysis of Pd-coated NPSS showed that the Pd particle size distribution was relatively narrow, with a mean particle size distribution of ~2 μm. On top of the 2 μm particles were smaller crystals of ~0.2 μm. Despite the apparent pinhole, the film was N_2-impermeable.

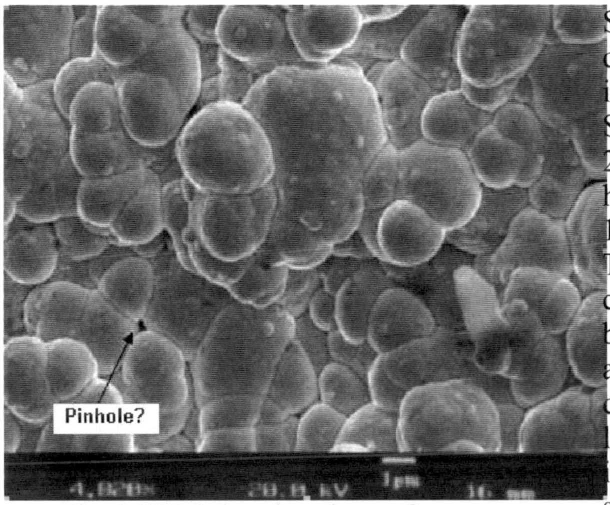

Fig. 1 SEM showing the surface topography of a Pd-coated NPSS circular disc after annealing in N_2 for 4 hours at 400°C and for 4 hours at 500°C

Silver deposition is much faster than Pd deposition, because nucleation of Pd or is slow; this explains the need for the Sn/Pd activation layer [25, 28]. Figures 2 and 3 show that the Pd-Ag crystals have an average crystallite diameter of ~ 1 μm. The surface looked like Ag/NPSS. The surface morphology should be controlled by Ag seeding at > 6 at. % Ag but film composition and morphology are very sensitive to plating bath chemistry [28]. It is possible that the lack of defined particle edges may have indicated incomplete diffusion between subsequent depositions, even under conditions consistent with Ma [29]. EDS analysis on bimetallic samples vs. annealing temperature was necessary.

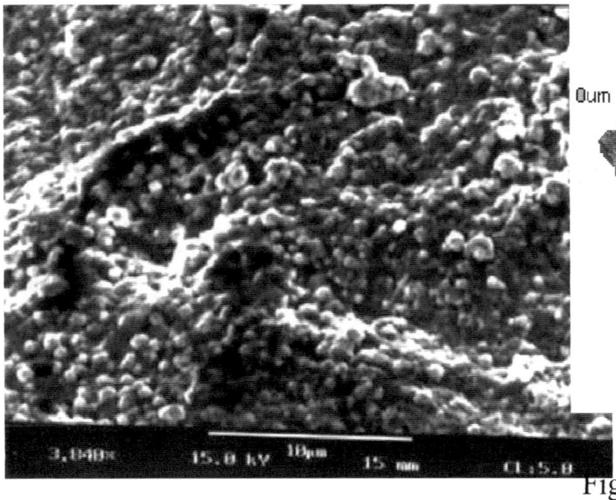

Fig. 2 SEM of codeposited Pd-Ag/NPSS after annealing in N_2 for 4 hours at 400°C and for 4 hours at 500°C

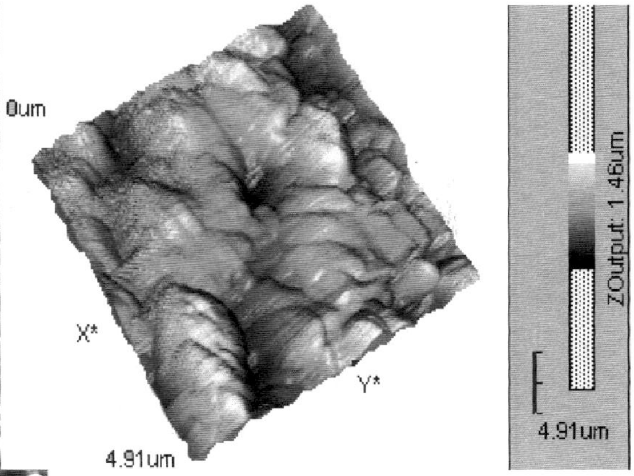

Fig. 3 - AFM image of codeposited Pd-Ag coated SS circular disc after annealing in N_2 for 4 hours each at 400°C and at 500°C

The key questions to address via SEM and EDS were whether the films had pinholes, and in the case of the bimetallic alloy films, whether the Pd and the Ag were properly alloyed. The latter issue is critical for matching the CTE's to minimize embrittlement. EDS mapping of some preliminary samples revealed pinholes in some preliminary samples. Once the aforementioned synthesis controls were applied, this problem was quickly solved. However, as indicated by the blue clusters in Fig. 5, 4 hours each at 500°C and 600°C in N_2 was insufficient for proper alloying of samples prepared via coating and diffusion (Fig. 4 for SEM). PSS-supported Pd-Ag alloys prepared via codeposition and then annealed at 600°C for 16 hours in N_2 showed that the Ag was uniformly dispersed, indicating proper alloy formation (Figs. 6 and 7). Lower temperatures and times were insufficient for proper alloy formation, in contradiction with Ma [29]. EDS revealed of the codeposited Pd-Ag/PSS weight percentages of Pd, Ag, and Fe, of 78.35, 20.04, and 1.35, respectively. This was close to the desired ratio of 75:25 for Pd and Ag. A small amount of Fe also diffused from the support into the Pd-Ag membrane.

Fig. 4 SEM of Pd-Ag alloy on NPSS via the coating and diffusion method. The substrate was annealed for 4 hours each at 500°C & 600°C in N_2.

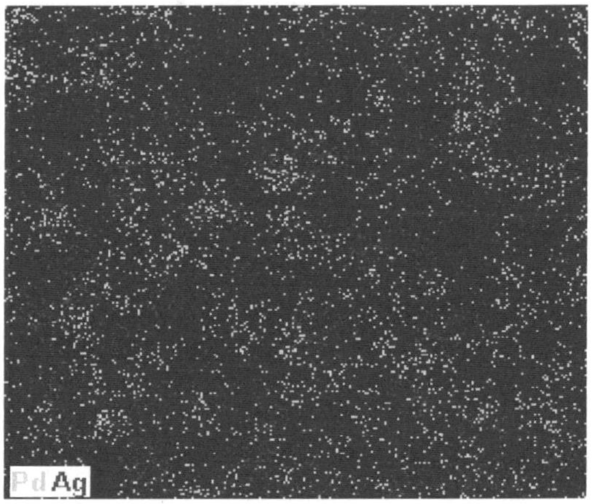

Fig. 5 EDS map of Pd and Ag on the Pd/Ag/NPSS substrate prepared using the coating and diffusion method shows that the Pd and Ag were not properly alloyed.

Fig. 6 SEM of codeposited Pd-Ag/PSS

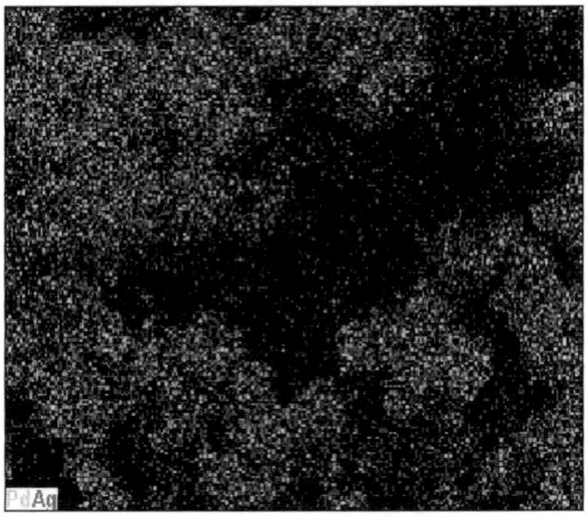

Fig. 7 Codeposited Pd-Ag/PSS EDS map

After realizing that at annealing temperatures below 500°C and times less than 10 hours were insufficient to form a proper Pd-Ag alloy, a semi-infinite slab model was used to estimate roughly the temperatures and times required to diffuse Ag uniformly to a desired extent into the Pd layer at a given Pd layer thickness. Prof. Ma e-mailed me the diffusivity data from [1], from which the annealing time required for Ag to diffuse uniformly in Pd-Ag layer in subsequent syntheses was calculated. Thin membranes yield both higher fluxes and better alloying [20]. Given that the diffusion times for diffusion distances of 1 μm at 873 K were realistic, all subsequent annealing was done at 873 K for 16 hours to assure that proper alloying was achieved. The benefit of the codeposition method over the coating & diffusion method is the shorter diffusion distance (≤ 1 μm).

Permeation Results

A N_2 permeance test was conducted at temperature ranges from 573 - 673 K to check for the impermeability of the prepared Pd-Ag purifier membrane. A very modest N_2 permeance of about 0.15 - 0.25 cm^3/min was detected at trans-membrane pressure difference of 250 kPa, much smaller than the H_2 flow rates (11 – 21 cm^3/min). A Sievert's Law plot (Fig. 8) was also made for the N_2 flux vs. the square root of the pressure difference between the retentate (P_r) and permeates side (P_p). This "law" held true for N_2, implying that N_2 dissociates into two nitrogen atoms from 573 - 673 K. Although N_2 flux was found to be temperature dependent, surprisingly the permeance was found to be independent of temperature. This may indicate pinhole diffusion or grain boundary diffusion, rather than interstitial diffusion, for N_2 permeance. Once the near impermeability of the prepared Pd-Ag membrane was confirmed with N_2, permeance with H_2 gas across the membrane was confirmed (Fig. 9).

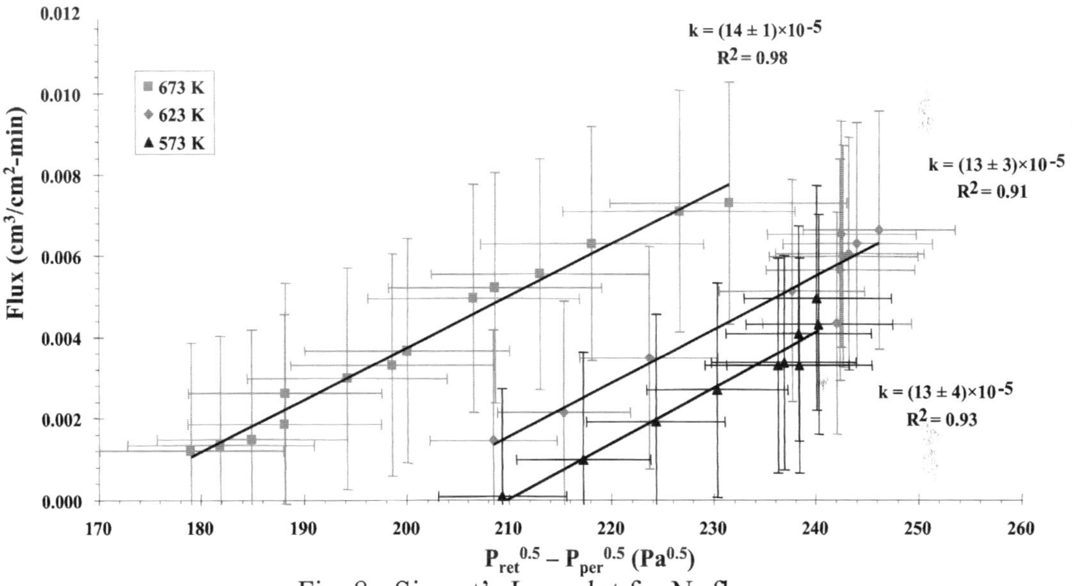

Fig. 8 - Sievert's Law plot for N_2 flux

The activation energy of the Pd-Ag membrane was found to be 9 ± 1 kJ/mol with a R^2 = 0.93, which was close to the previously reported values for Pd or Pd alloy membranes. A comparison between the derived experimental activation energy value for the Pd-Ag/PSS alloy membrane and those of other researchers is shown in Table 3.

The H_2 permeability and permeance can also be compared to work done by other researchers in Table 4. Figure 10 compares the H_2 and N_2 fluxes and confirms the relative impermeability of the Pd-Ag membrane to non-H_2 gases. Long-term stability of the Pd-Ag membrane was verified in Figure 11.

Table 3: Activation Energies for H_2 Permeation

Support	Membrane thickness	Temp (K)	E_d (kJ/mol)	Reference
Pd/Ag/PSS	48*	623-823	9 ± 1	This work
Pd/PSS	19-28	673-1273	16.38	Mardilovich [22]
Pd discs	1000	623-1173	13.41	Rothenberger [14]
Pd/disk	486-762	300-709	11.81	Hurlbert [30]
Pd/Ag/ceramic	0.3	673-993	23	Lin [31]
Pd/disc	800-2025	500-900	12.81	Holleck [32]

*This was the membrane thickness on both sides (inside and outside) of the PSS membrane. Pd alloy thickness on one side of the membrane would be 24 microns. All the other work reported is for Pd layer only on the outside of the membrane substrate.

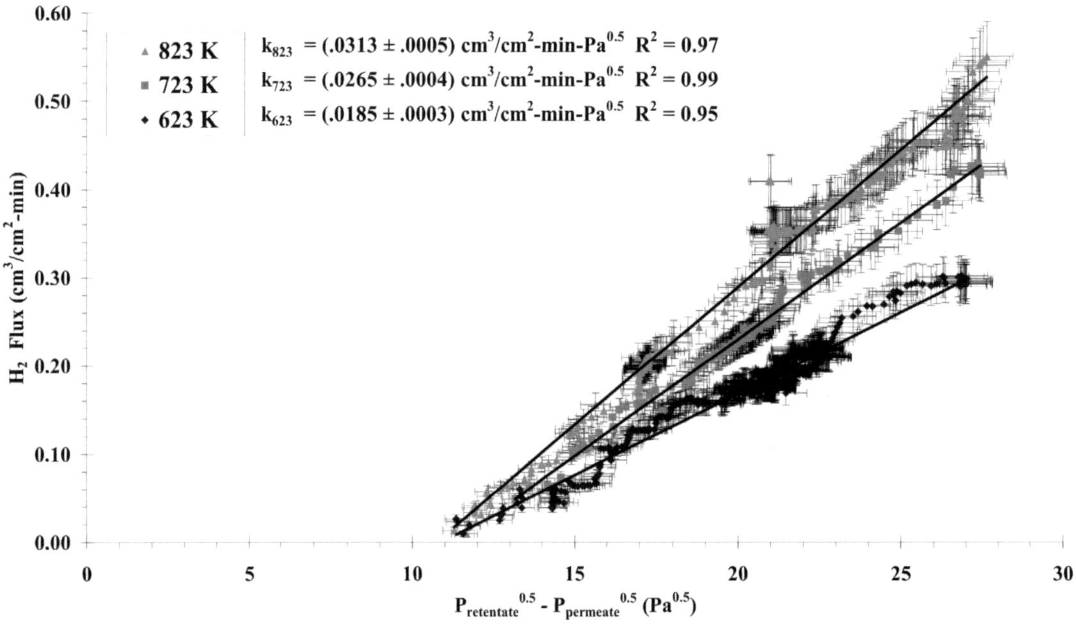

Fig. 9 - Sievert's Law plot for H_2 permeance

Table 4: Comparison of H_2 Permeances and Permeabilities

T	k (cm^3/cm^2-min-$Pa^{0.5}$)	K (cm^3/cm-min-$Pa^{0.5}$)	Ref.
623	(1.85 ± 0.03)E-02	(8.9 ± 0.2)E-05	This work
623	4.33E-02	9.61E-05	[20]
623	4.15E-02	9.96E-05	[22]
673	3.62E-02	7.78E-05	[20]
673	3.75E-02	9.00E-05	[22]
723	(2.65 ± 0.04)E-02	(12.8 ± 0.2)E-05	This work
773	4.30E-02	8.60E-05	[26]
823	(3.13 ± 0.05)E-02	(15.1 ± 0.3)E-05	This work
823	8.70E-03	2.12E-04	[24]
850	1.08E-01	1.72E-05	[2]
1023	3.56E-03	5.56E-05	[24]

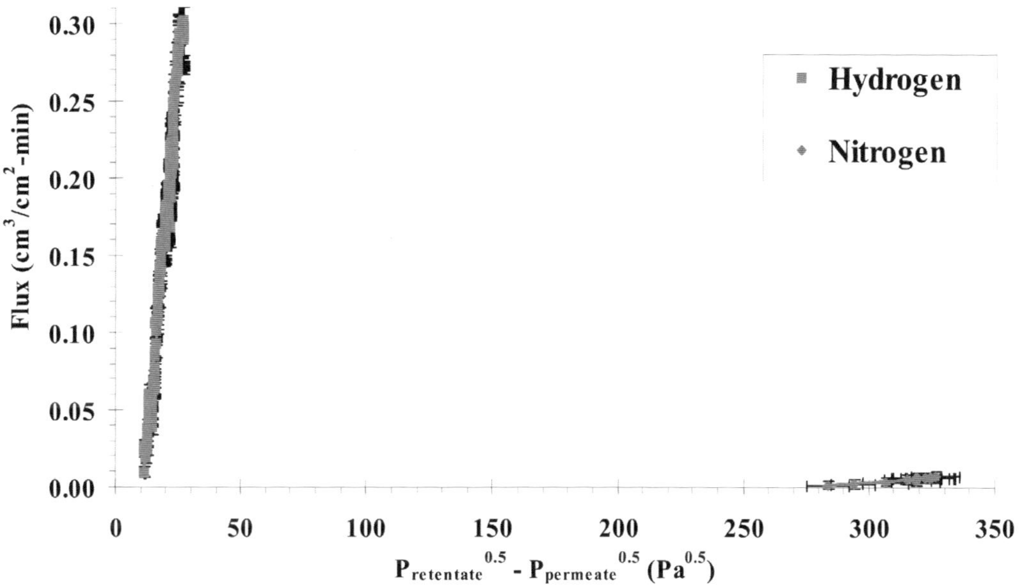

Fig. 10 - H_2 Flux vs. N_2 Flux at 623 K

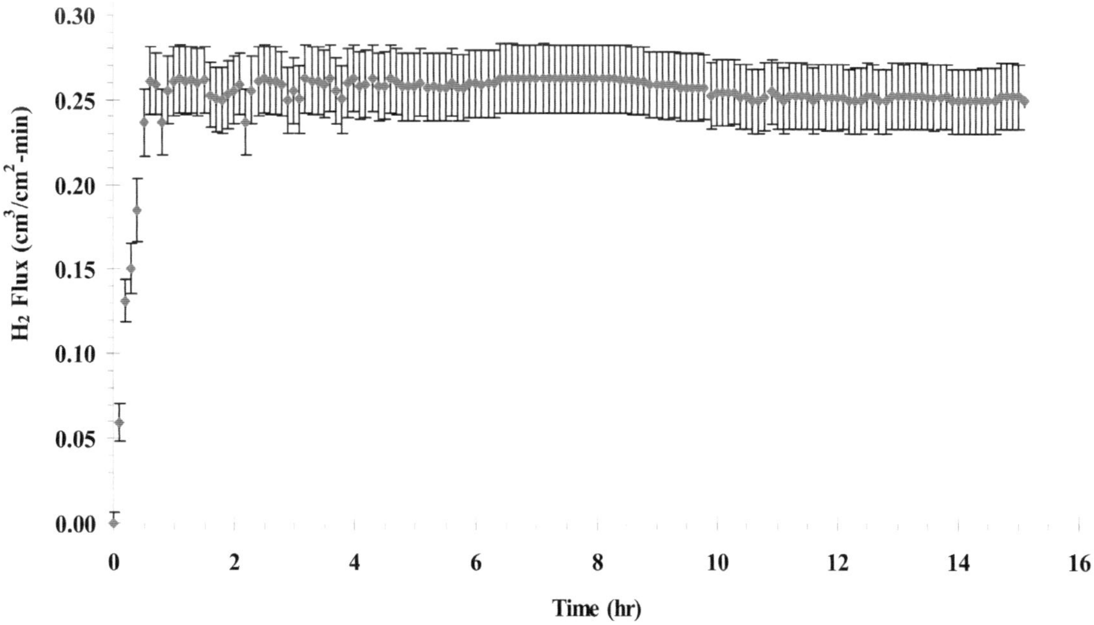

Fig. 11 - H_2 permeance stability at 623 K

Chapter 4: Conclusions

1. The proper conditions for co-deposition of Pd and Ag from a mixed plating bath were identified and verified: [Metal] = 10 mM, [N_2H_4] = 10 mM, [NH_4OH] = 3 M, pH = 10.4 and T= 60°C. Sufficient interdiffusion of the two metals to form a proper alloy requires annealing at 600°C for 16 hours in N_2.
2. SEM-EDS and AFM results concluded that a relatively smooth layer of 79Pd-21Ag alloy was developed on the porous stainless steel substrate with an average Pd-Ag crystal diameter of 0.5 μm with a height variation of about 0.2 μm.

3. EDS revealed that only trace levels of support Fe diffused into the Pd-Ag layers.
4. Very minor N_2 permeation could be detected at temperatures of 573-673 K. The N_2 permeance was found to be constant at 1.3×10^{-4} cm^3/cm^2-min-$Pa^{0.5}$. The magnitude of the N_2 leak was comparable to those reported by others [29]. Impermeability of the prepared membrane needs to be tested with He gas. The results do not support N_2 diffusing interstitially just like H_2 through Pd-Ag.
5. Membranes composed of thin film of Pd (48 μm) and Pd/Ag deposited on porous stainless steel via electroless plating have been evaluated at temperatures of 623-823 K with H_2 permeability from $(9\text{-}15)\times10^{-5}$ cm^3/cm-min-$Pa^{0.5}$.
6. The activation energy of the Pd-Ag membrane was found to be 9 ± 1 kJ/mol, which was close to the previously reported values for Pd or Pd/Alloy membranes.
7. The reproducibility and long time stability of the H_2 permeation was confirmed.
8. In order to increase the H_2 flux, a preparation technique to develop a nanoscale Pd layer will be the focus of future work.

References

1. Hou, K.; Hughes, R., *J. Membrane Sci.*, **214**, 2003, 43-55.
2. Yeung, K.L.; Christiansen, S.; Varma, A., *J. Membrane Sci.*, **159**, 1999, 107-122.
3. O'Brien, J.; Hughes, R.; Hisek, J., *Surface and Coatings Technology*, **142-144**, 2001, 253-259.
4. Kikuchi, E.; Nemoto, Y.; Kajiwara, M.; Uemiya, S.; Kojima, T., *Catalysis Today*, **56**, 2002, 75-81.
5. Moron, F.; Pina, M.P.; Urriolabeitia, E.; Menendez, M.; and Santamaria, J., *Desalination*, **147**, 2002, 425-431.
6. Li, A.; Xiong, G.; Finghua G., Zheng, L., *J. Membrane Sci.*, **110**, 1996, 257-260.
7. Chen, H.I.; Shiau, J.D.; Chu, C.Y.; Huang, T.-C., *Separation and Purification Technology*, **32**, 2003, 247-254.
8. Jun, C.S.; Lee, K.H., *J. Membrane Sci.*, **157**, 1999, 107-115.
9. Shu, J.; Grandjean, B.P.A.; Kaliaguine, S., *Can. J. Chem. Eng.*, **75**, 1997, 712.
10. Nam, S.E.; Lee, K.H., *J. Membrane Sci.*, **170**, 2000, 91-99.
11. Pan, X.L.; Stroh, N.; Brunner, H.; Xiong, G.X.; Sheng, S.S., *Separation and Purification Technology*, **32**, 2003, 265-270.
12. Shu, J.; Grandjean, B.P.; Neste, A.V.; Kaliaguine, S., *Can. J. Chem. Eng.*, **69**, 1991, 1036-1074.
13. Rothenberger, K. S.; Howard, B. H.; Killmeyer, R. P.; Cugini, A. V.; Morreale, B. D.; Enick, R. M.; Bustamante, F., *J. Membrane Sci.*, **241(2)**, 2004, 207-218.
14. Rothenberger, K.S., *J. Membrane Sci.*, **212**, 2003, 87-97.
15. Shu, J.; Grandjean, B.; Neste, A.V.; Kaliaguine, S.; Dalmon, A.J., *J. Chem. Soc.*, **54**, 1996, 2475.
16. Kikuchi, E.; Nemoto, Y.; Kajiwara, M.; Uemiya, S.; Kojima, T., *Catalysis Today*, **56**, 2002, 75-81.
17. Meenan, B.J.; Brown, N.M.D.; Wilson, J.W., *Applied Surface Science*, **74**, 1994, 221-233.
18. Yan, S.; Maeda, H.; Kusakabe, K.; Morooka, S., *Ind. Eng. Chem. Res.*, **33**, 1994, 616.
19. Uemiya, S.; Sato, N.; Ando, H.; Kikuchi, E., *Ind. Eng. Chem. Res.*, **67**, 1991, 585.
20. Ma, Y.H.; Mardilovich, I.P.; Engwall, E.E., *Annual of the New York Academy of Science*, **984**, 2003, 346.

21. Lin, Y.M.; Lee, G.L.; Rei, M.H., *Catalysis Today*, **44**, 1998, 343.
22. Mardilovich, P.P.; She, Y.; Ma, Y.H.; Rei, M.H., *AIChE Journal*, **44(2)**, 1998, 310-322.
23. Tosti, S.; Bettinali, L.; Violante, V.; "Rolled Thin Pd and Pd–Ag Membranes for Hydrogen Separation and Recovery," *Int. J. Hydrogen Energy*, **25**, 2000, 319.
24. Jarosch, K.; Lasa, H.; "Permeability and Selectivity, and Testing of Hydrogen Diffusion Membranes Suitable for Use in Steam Reforming," *Ind. Eng. Chem. Res.*, **40,** 2001, 5391.
25. Shu, J.; Grandjean, B.P.A.; Ghali, E.; Kaliaguine, S.; "Simultaneous Deposition of Pd and Ag on Porous Stainless Steel by Electroless Plating", *J. Membrane Sci.* **77**, 1993, 181.
26. Shu, J.; Adnot, A.; Grandjean, B.P.A.; Kaliaguine, S.; "Structurally Stable Composite Pd/Ag Alloy Membranes: Introduction of a Diffusion Barrier", *Thin Solid Films* **286,** 1996, 72.
27. Ward, T.L.; Dao, T.; "Model of Hydrogen Permeation Behavior in Palladium Membranes," *J. Membrane Sci.*, **153**, 1999, 211-231.
28. Yeung, K.L.; Cheng, T.S.; "Palladium Silver Composite Membranes by Electroless Plating Technique," *J. Membrane Sci.*, **158**, 1999, 127-141.
29. Ma, Y.H.; Mardilovich, I.P.; Engwall, E.E.; "Composite Pd/PSS Membranes for Production of Pure Hydrogen and High Pressure CO_2 for Sequestration" (Oral Paper 77a), *Spring 2005 AIChE Conference*, Atlanta, GA, April 10-14, 2005.
30. Hurlbert, R.C.; Konecny, J.O.; "Diffusion of Hydrogen through Palladium" *J. Chem. Phys.*, **34 (2)**, 1961, 655-658.
31. Lin, Y.S.; Jayaraman, V.; "Synthesis and Hydrogen Permeation Properties of Ultrathin Palladium-Silver Alloy Membranes", *J. Membrane Sci.*, **104**, 1995, 251.
32. Holleck, G.L.; "Diffusion and Solubility of Hydrogen in Palladium and Palladium-Silver Alloys", *J. Phys. Chem.*, **74 (3)**, 1970, 503-511.